BIOMEDICAL RESEARCH

International Perspectives

VIBS

Volume 186

a volume in
Values in Bioethics
ViB
Matti Häyry and Tuija Takala, Editors

ETHICS IN BIOMEDICAL RESEARCH

International Perspectives

Edited by

Matti Häyry, Tuija Takala, and

Peter Herissone-Kelly

With a Foreword by

Alexander Morgan Capron

Rodopi

Amsterdam - New York, NY 2007

Cover image courtesy Su-Chun Zhang

Cover Design: Studio Pollmann

The paper on which this book is printed meets the requirements of "ISO 9706:1994, Information and documentation - Paper for documents - Requirements for permanence".

ISBN-13: 978-90-420-2179-2

Printed in the Netherlands

CONTENTS

FOREWORD

WHY BOTHER WITH ETHICS IN BIOMEDICAL RESEARCH?

Biomedical research—from pioneering investigations of molecular and cellular functions to Phase III clinical trials of new medicines—is inherently interesting, fascinating at a scientific level. It is of immense importance to human life and well-being. These are reasons enough to explain why scholars, policy-makers and members of the public have long been concerned that such research be conducted ethically. But the reasons go much deeper. We worry about the ethics of biomedical research because we can never entirely escape the specter of the Nazi concentration camps, Unit 731, the Tuskegee syphilis study, the post-war radiation experiments, and countless other instances where skilled medical scientists allowed their thirst for knowledge and power to cause harm—often, unspeakable harm—to their fellow human beings.

Typically, those who suffered most were members of vulnerable groups, a factor that makes this abuse more inexcusable. Failure to protect society's weak condemns all of society. More understandable, they were powerless to resist and often unaware that they were being used in research. Frequently, these groups were vulnerable because they were marginalized by their ethnicity, their disabilities, and their poverty. But we can never feel confident that the harms that arise in research will only happen to "them," much less that they will only happen in another time or another culture. We have only to open the daily paper to read of prestigious medical schools whose research programs have been halted because of faulty ethical review procedures or physicians who have enrolled patients in clinical trials without disclosing the big bonuses they have been promised for doing so.

Bothered though we may be by such risks, we are not inclined to forgo biomedical research, for its fruits are far too precious. The prospect that researchers will decode the human genome, forestall the aging process, or conquer dread diseases has led to ever-increasing allocations of public funds for biomedicine. The lure of more pills to solve more problems suffered by human beings has driven private spending by pharmaceutical and biotech companies to even greater heights. Yet the rapid growth of biomedical research and its movement away from academic centers and from developed countries into the developing world only reinforce the conviction that ethical principles, rules, and processes need attention too, now more than ever.

Consequently, nothing could be timelier than the present collection of essays in which a truly international group of scholars directly addresses the hopes and fears that arise as biomedical research becomes ever more global. While such research could prove a strong force for development, the authors raise essential questions about the meaning of the concept of "development," the limits of technology, and the new forms of vulnerability that flow from the huge international health inequalities that now exist.

Will the present system produce a research agenda responsive to the health needs and respectful of the cultural values of all potential participants? What obligations do researchers owe to their subjects and to others? Conversely, do we, as citizens and especially as patients, owe any duties to biomedical research, whose fruits we are so willing to enjoy?

As these essays demonstrate, a careful look at the ethics of biomedical research inevitably involves more than the familiar principles of autonomy and beneficence, for notions of justice, of power, and of accountability are also crucial components of any adequate analysis. For anyone interested in understanding the most important contemporary ethical issues in biomedical research, this volume is essential reading.

Alexander Morgan Capron, Director
Department of Ethics, Trade, Human Rights and Health Law
World Health Organization.

Genève, Switzerland
August 2005

PREFACE

On 21–23 August 2003, an international symposium was held in Helsinki on Ethics in Biomedical Research. Most of the chapters of this book were presented in that symposium by the then Directors of the International Association of Bioethics. The editors wish to thank the organizers of the event, Mika Tirronen, Laura Walin, and Liisa Teräsvuori; the President of the Symposium, Academician P. Helena Mäkelä; its Vice President, Sakari Karjalainen; the speakers, panelists, and participants of the event; and the following sponsors for their contributions:

- Life 2000 Research Programme of the Academy of Finland and the National Technology Agency, Tekes
- Helsinki Biomedical Graduate School
- The Finnish Medical Society Duodecim
- Turku Graduate School of Biomedical Sciences
- Drug Discovery Graduate School
- Department of Social and Moral Philosophy, University of Helsinki
- Centre for Professional Ethics, University of Central Lancashire
- Faculty of Health, University of Central Lancashire
- Centre for Social Ethics and Policy, The University of Manchester
- Institute of Medicine, Law and Bioethics, The University of Manchester
- School of Law, The University of Manchester
- International Association of Bioethics

The editors are also grateful to Elizabeth D. Boepple, who prepared the final manuscript with great skill and dedication.

Matti Häyry
Tuija Takala
Peter Herissone-Kelly

Manchester and Preston

Introduction

DEVELOPMENT, RESEARCH, AND VULNERABILITY

Matti Häyry, Tuija Takala, and Peter Herissone-Kelly

1. The Internationality of Research Ethics

The ethical assessment and regulation of biomedical research is a genuinely international and global matter. Wherever conducted, the ramifications and implications of such research have the potential to affect people all over the world. Viewing work in affluent or technologically advanced regions against the background of its impact on other countries makes sense. Evaluating scientific work performed in any country in terms of its consequences for defenseless sections of the population is imperative.

The authors of this book address three major aspects of internationality in research ethics, namely global development, scientific advances, and vulnerability. Their questions and answers, practical and theoretical, go straight to the heart of the issue, and in many ways provide guidance to decision makers in the field.

2. Facts, Fears, and Hopes Concerning Development

The first part of the book deals with questions of development.

In chapter one, Solomon R. Benatar sets the practical scene by outlining some of the most glaring disparities and conflicts in international collaborative clinical research. Current pharmaceutical development aims more at products that will produce high profits than at treatments that can prevent or cure the most prevalent diseases in the world. Research is, in many cases, conducted in the developing world, while the affluent nations enjoy the benefits of its results. We can find the reasons for this in the phenomenon of globalization in its many forms, but Benatar believes that change is possible. By refocusing our efforts on prevention of infectious diseases, fair sharing of benefits, and the universal standard of care, we could reverse the trend that widens the gap between nations, regions, and populations.

In chapter two, Matti Häyry sketches a theoretical background for the discussion by presenting an interpretation of the history of bioethics. According to him, we can view bioethics as an enterprise that aims at the protection of the dignity and autonomy of vulnerable individuals and groups. This, he claims, is uncontroversial, although controversies can arise when different people define dignity, autonomy, and vulnerability in different ways. Häyry sees in the development of Western bioethics two phases: first a period of anti-paternalism, then a new era in which the empowerment of the professionals is more to the fore. Despite the contrast between these stages, both stress the dignity and autonomy of the vulnerable, the first as the protection of patients and research subjects from authoritarian physicians and scientists, the second as the salvation of humankind from curable ailments.

In chapter three, Florencia Luna turns our attention to the limited applicability of sophisticated technologies in non-industrialized countries. She starts by questioning the epithet "developing" in the context of nations whose economies are stagnant or deteriorating, and points out that economic, social, and cultural factors account for considerable diversity in the needs of different populations. In addition to differences among nations, affluence and poverty co-exist within most nation states and regions. Luna argues that to improve conditions globally and locally, we should direct more attention to public health considerations than to individualistic measures. This means, among other things, that we should re-gear existing technologies towards addressing practical issues in resource-poor countries, and that we should develop new technologies by having in mind the needs of these countries.

In chapter four, Leo Pessini and Leonard M. Martin provide an optimistic view of the development of research ethics in one so-called developing nation, Brazil. They note that the regulation of scientific activities in their country has benefited considerably from an application of concepts and principles prevalent in North American and European moral debates, notably autonomy, nonmaleficence, beneficence, justice, and dignity. These notions provide an ethical, as opposed to a purely legal or political, basis for the assessment of biomedical research. At the same time, Pessini and Martin argue that the Brazilian model could ideally contain valuable lessons for the more industrialized regions. They emphasize the superiority of practical guidelines and social control as responses to scientific advances over abstract ideals and the self-control of those conducting the research.

In chapter five, Angela Amondi Wasunna delineates some of the responsibilities project sponsors have toward studied and wider populations in developing countries. She draws a distinction between research that responds to a health need considered crucial in the studied community, and work that responds to a lower profile need. The first option is ideal. Although the rank order set by a nation or a population, or implied in its tradition, can be difficult to identify, Wasunna stresses the obligation of the scientists and the sponsors to engage in a dialogue with the community to agree on the needs that we should prioritize over others. The second alternative is more questionable, but Wasunna believes that we need not rule it out entirely, as long as the researchers' work benefits the studied community in a way to which they have agreed after open and equal negotiation.

3. Research and Technology as Answers and as Questions

The second part of the book is concerned with the ethics of genetics in its many forms.

In chapter six, Darryl R. J. Macer considers the case of controlling infectious diseases like dengue fever, malaria, and Chagas disease by introducing genetically modified mosquitoes into the environment. Although most people would agree that the practice does not have intrinsic moral problems, it does raise questions concerning the safety of the human beings affected and the balance of the environment in which the interventions are made. Since the impact knows no national borders, Macer argues that we should develop a universal standard to determine acceptable levels of risk. He also suggests that communication between researchers and the local population plays a significant role in the assessment of experiments, and that community involvement, group consent, and negotiations between the primary target population and neighboring countries are strongly advisable.

In chapter seven, Michael Parker, Richard Ashcroft, Andrew Wilkie, and Alastair Kent move to the opposite end of the continuum, in terms of the number of people concerned, and tackle the ethical issues of rare genetic disorders. Investigations into these can take time, and they often more closely resemble research than do other clinical examinations. Their classification as research can be beneficial in that, under this label, gaining funding and getting results published in professional journals might be easier. On the other hand, as research, these activities have to be assessed by ethics committees; the singularity—non-generalizability—of the cases, among other things, can make meeting the standards of those committees difficult. Parker and co-authors conclude that, in the absence of a well-defined distinction between the categories, we should view these activities as clinical investigations unless the researcher begins to search for affected individuals outside the family of the original patient.

In chapter eight, Rosemarie Tong describes the fascination with which people in the United States discuss matters of reproduction and the prolongation of life by technological advances. She first considers questions of the value of human life, and biomedical research on fetuses, embryos, and pre-embryos, and then outlines the debate over therapeutic and reproductive cloning. One salient feature in these exchanges is, she contends, the tension between strong beliefs concerning the worth of beginning lives and an underlying trust in a better world where people are healthier and live longer due to new forms of research. Especially in her section on ectogenesis—gestating human beings in an artificial womb—Tong exposes the controversies that arise when the reproductive autonomy of women is in conflict with the ideology that every human life is worth protecting from its earliest developmental stages.

In chapter nine, Bart Hansen and Paul Schotsmans investigate ethical issues regarding stem cell research and reproductive cloning from a Christian viewpoint. They argue that a view of human beings as "created co-creators" instead of mere keepers of God's nature can more accurately grasp the essence of humanity. This image implies an imperative to study God's creation and to be involved in its development in good and right directions. Some Christian theologians have judged human reproductive cloning and experiments on embryos produced solely for the purpose of research as immoral. Hansen and Schotsmans maintain that this outlook is unnecessarily restrictive. They believe that scientists as created co-creators are entitled to improve the lot of humanity in ways that include therapeutic cloning and stem cell research.

In chapter ten, Boris Yudin discusses the similarities and dissimilarities between different ethical outlooks. He emphasizes the responsibilities that we have in the face of technological change. The ethics of biotechnology, as he sees it, has two value orientations, and these can pull our moral considerations in opposite directions. The first is preservation. Many theorists stress the significance of care and precaution, and the potential untoward impact of technologies on society and the natural environment. The second is change. Another group of ethicists and scientists see technological advances as an instrument for us to realize our hopes and dreams of a better life. Yudin notes that both views are defensible. Human beings have arguably interfered with natural processes for a long time, and the results have not always been bad. Yet the issue of preservation is also noteworthy due to our ever-increasing capacity to occasion change.

4. Vulnerability, Power, and Responsibility

The third part of the book takes up issues of human vulnerability in biomedical and health research.

In chapter eleven, Leslie P. Francis, Margaret P. Battin, Jeffrey R. Botkin, Jay A. Jacobson, and Charles B. Smith draw attention to the role of indirect participants in the study of infectious diseases. In most investigations of communicable diseases, researchers have seen informed consent of the study subjects—direct participants—as the top priority. They have traditionally paid less heed to risks of contagion to the family and community members of those studied, although examples of this kind of attention also exist. Research on immunization, for instance, can be as dangerous to the people around the direct subjects as to the subjects themselves. Francis and co-authors argue that indirect participants should enjoy the same level of protection as the persons studied, although investigators should exercise caution when deciding how far beyond the immediate contacts to extend the sphere of protection.

In chapter twelve, Leonardo D. de Castro examines the claim that citizens of modern nations have a duty to serve as subjects of biomedical research. He recognizes the importance of such research and concedes that an initial case exists for believing that we share such a moral obligation. Further analysis shows that general justifications fail to support it. Analogies to civic duties are ultimately unhelpful since even if we agreed that an obligation to produce medical benefits to other people existed, people could discharge their obligation in ways other than participation as research subjects. Our duties to promote social fairness and to prevent harm are subject to similar objections. De Castro argues that an obligation might bind people who belong to some recognizably vulnerable groups, such as emergency patients, children, and pregnant women. Even in these cases, though, considerations of autonomy should prevent us from blaming people if they fail to act in ways we view as moral.

In chapter thirteen, Elma Lourdes Campos Pavone Zoboli creates an analytical framework for assessing vulnerability in biomedical research. Her starting point is in a historical model originally produced to evaluate the different policies of protecting and empowering persons with HIV/AIDS. These policies started with an individualistic emphasis, but soon evolved to have social and political dimensions. Zoboli maintains that, in the same way, we should acknowledge the vulnerability of research subjects on all three levels. The dimension of informed consent, currently stressed, is only one such consideration. Research committees and regulation are also crucial social controls. In the political arena, we should make sure that the subject populations have been selected fairly and without coercion, exploitation, or other undesirable external pressures.

In chapter fourteen, Maria Patrão do Céu Neves discusses the relationship between power and duty in the political control of science and technology. Drawing on the works of Emmanuel Lévinas and Hans Jonas, she outlines three types of power that characterize modern biomedicine, namely therapeutic, economic, and social. Holding that those who can influence the lives of others have an obligation to protect them, Patrão Neves argues that all three kinds of power imply duties to researchers and to society as a whole. She says that to be content with the "ethic of rights," in which the autonomy of individuals is paramount, is not sufficient. What we need, according to her, is an "ethic of duties." This would help us, in a world where globalization is ever increasing, to develop and cherish attitudes of care and responsibility, which in their turn would make possible an ability to encounter human vulnerability, in others and ourselves.

In chapter fifteen, Frank J. Leavitt considers some concrete alternatives to biomedical research as an attempt to make the world a better place by bringing health and longevity to greater numbers of people. Placing prevention firmly before cure, he argues that we cannot address today's most important global risks to health by high-tech medicine, but we could reduce these risks by taking practical, natural measures. As examples of such natural measures, he discusses education for disease prevention, the kitchen garden model for producing nutritious vegetables, and individual health autonomy, understood as the responsibility individuals have to safeguard their health, instead of relying on medical measures for recovery after compromising it. Although obstacles stand in the way, Leavitt believes that it would be useful to direct some of the current research resources into the study of simple pragmatic solutions like these.

5. The Role of International Bioethics

We have much left to be done before we reach a situation where we satisfactorily protect, cherish, and promote the dignity and autonomy of everyone in the world. International bioethics has a significant role to play in this process, as evinced by these contributions. The answers to some of the questions posed in this book are not easy to discover. We hope that we have asked some of the right questions and clarified their ramifications and implications. Clarification of the questions, we believe, is a crucial first step toward finding the answers.

Part One

FACTS, FEARS, AND HOPES CONCERNING DEVELOPMENT

One

NEW PERSPECTIVES ON INTERNATIONAL RESEARCH ETHICS

Solomon R. Benatar

1. Themes and Background

In our modern world, we have great faith in scientific progress, economic growth, and human rights as the essential ingredients for improving life for all. Sadly, we live in a world in which despite great progress on all these fronts, life is deteriorating for most of the world's people. Widening disparities in wealth and health, and the emergence of new infectious diseases and environmental degradation pose threats to the lives of all, including those who have benefited most from progress. Many of our attitudes and actions will need to change radically to ameliorate this situation.

This paper addresses the conduct of international collaborative clinical research. I will review trends in how we conduct such research, and I will discuss reasons why investigators persist in their current emphasis that favors research on diseases that account for only 10 percent of the global burden of disease. I conclude by exploring new perspectives that could contribute to promoting much needed progress in health worldwide.

Issues of relevance to international research ethics include:

(1) The features of a globalizing world that influence and shape the research agenda;
(2) Emergence of new infectious diseases, especially HIV/AIDS, as a signs of global instability and global threats;
(3) HIV/AIDS as influential in changing ethical attitudes towards research on vulnerable subjects;
(4) Some persisting controversies in international clinical research;
(5) The standard of care for participants in international research ethics; and the
(6) Potential for new perspectives on the ethics of international research to contribute to much needed progress in human health worldwide.

2. How International Collaborative Clinical Research Is Conducted: Trends

Clinical research has become a growth industry.[1] Between 1995 and 1998, the number of drugs tested pre-clinically in animals increased by 27 percent. Enrollment of subjects in multi-center trials increased 221 percent from the 1980s to the 1990s. The number of overseas investigators (developing country partners) increased by 444 percent between 1992 and 1998. This growth in clinical research and in collaboration with colleagues in the developing world results from several factors. First, investigators have a greater dependence on the pharmaceutical industry for funding, and most universities in the developed and developing world receive inadequate funding to conduct research without the support of industry. Second, because industry has an agenda, and pharmaceuticals have high market value, research is increasingly driven by the search for drugs with high use potential. Third, to rapidly obtain definitive results, large trials must be conducted to register and market new drugs as quickly as possible. Diseases that interest the pharmaceutical industry, such as hypertension, diabetes, pneumonia, asthma, chronic obstructive lung disease, arthritis, osteoporosis, and depression are all manifest world wide, including the developing world. Fourth, access to developing-country subjects is often much easier and cheaper than in the developed world.

A few statistics about scientific research reveal that the research endeavor is not value free. Two thirds of the United States Government's expenditure on research and development is on military research.[2] Ninety percent of global funding for medical research, about $70 billion per year, is spent on diseases causing only 10 percent of the global burden of disease.[3] Of 1,393 new chemical entities marketed between 1975 and 1999, only sixteen were for tropical diseases and tuberculosis that account for enormous mortality and morbidity.[4] These data are not surprising given that the private sector and the pharmaceutical industry sponsor about 50 percent of medical research. Such observations suggest that the research agenda pursued by scientists is less determined by the need for knowledge and how to apply knowledge than by economic forces that drive and support research endeavors. Emphasis on military research, with significant neglect of diseases that afflict billions of people living in abject misery, reflects a value system that devalues the lives of more than half the world's population.

We have much unused knowledge, implementation of which could save many lives. This suggests that researchers' interests can outweigh the best interests of subjects. We appear to value acquiring new knowledge more highly than applying existing knowledge. In answer to those who urge more research because more lives will be saved, we note that saving lives in poor countries is not dependent predominantly on medical research—more than 27,000 children under five years (10.1 million children annually) old die daily from hunger and preventable diseases. Such deaths are not due to lack of either knowledge or resources. Fifty billion dollars, which amounts to only 0.2 percent of the gross domestic product of affluent countries, could prevent up to 50 percent of premature deaths.[5] By comparison, the Iraq war cost over $100 billion, and the World Health Organization's annual budget is less that $2 billion.

3. Why Do We Persist with the Current Emphasis?

To understand the forces influencing prevailing research agendas, we must appreciate that modern clinical research does not take place in a vacuum, but pursued within the context of a world increasingly influenced by powerful forces. Globalization, a widely used term meaning different things to different people, has generally accepted essential features such as changing perceptions of time and physical space, and diffusion of ideas, culture, and values on a global scale. While economic globalization is the most commonly used context, globalization is a complex concept that goes beyond economics to include social, cultural, and ecological dimensions. Globalization is not a new phenomenon, but the outcome of a long interwoven economic and political history, involving a wide range of actors, with beneficial and adverse effects on human well-being, although critics can argue that, like population growth, the adverse effects of globalization are now becoming starkly apparent. Gillian Walt has described globalization as being characterized by five, often conflicting, themes:

(1) Economic transformation (financial volatility, marginalization, labor insecurity);
(2) New trade regimes (with winners and losers);
(3) A growing poverty gap (with rising health inequalities);
(4) An electronic revolution (that separates the "knows" and the "know-nots"); and
(5) New forms of governance (that include a proliferating spectrum of non-state actors).[6]

Economic globalization has been ostensibly spearheaded by a few hundred corporate giants, the development of earth-spanning technologies, products that can be produced anywhere and sold everywhere, and the spreading of credit through pervasively penetrating global channels of communication. Four interconnecting and integrated transnational commodity chains have been described:

(1) A global cultural bazaar (the entertainment industry and communication networks);
(2) A global shopping mall (through which tobacco, food, and clothes are promoted by aggressive advertising, distribution, and marketing);
(3) A global work place of factories, law offices, hospitals, restaurants (facilitated by population growth, mass production, automation, and cheap labor); and

(4) A set of global financial networks (that include speculative devices for repackaging and reselling money in what is increasingly labeled a "casino economy").[7]

Positive manifestations of progress associated with globalization include advances in science and technology, increased longevity, enhanced economic growth, greater freedom and prosperity for many, improvements in the speed and cost of communications and transport, and popularization of the concept of human rights.

Negative effects of globalization include widening economic disparities between rich and poor within and between nations, and increases in absolute and relative poverty. At the beginning of the twentieth century, the wealthiest 20 percent of the world's population were nine times richer than the poorest 20 percent. This ratio has grown progressively to thirty times by 1960, sixty times by 1990, and to over seventy-four times by 1997. World debt grew from $0.5 trillion in 1980, to $1.9 trillion in 1994, and $ 2.2 trillion in 1997. The way in which debt is created and sustained and its relationship to the arms trade have been major factors in perpetuating and intensifying poverty and ill health.[8] Most countries the World Bank required to pursue structural adjustment programs are in greater debt than ever before. Third-world debt, although accounting for only a small proportion of total world debt, has reached exorbitant levels relative to income in the third world. These countries cannot repay their debts.[9]

The evolving complex web of material, institutional, and ideological forces, and the power of massive multinational corporations in a globalizing world has profound implications for the accumulation of capital and for the way in which resources are controlled. In 1970, 70 percent of all money that changed hands on a daily basis was payment for work, while speculative financial transactions accounted for only 30 percent. By 1998, when daily speculative exchanges amounted to $1,500 billion daily, these proportions had changed to 5 percent and 95 percent respectively.[10] Such a striking shift in the distribution of money arguably reflects devaluation of the lives and work of most people in the world.

The shift in locus of economic power from the nation state to global corporations is altering the balance of power in the world. Boundaries between states and between foreign and domestic policies are being blurred, in the process, undermining small states' control over their economies and threatening their ability to provide for their citizens. Economic disparities have become so marked and adverse effects so apparent, that a significant degree of incompatibility has arisen between neo-liberal economic policies and the goals of democracy.[11] In addition to progressive widening of the economic divide between nations, and growing external control by moneylenders over the economies of small countries through debt, trade, and markets increasingly global, other powerful forces, for example, feminization of labor, more part-time employment, and exploitation of cheap labor in developing countries, are creating new patterns of inclusion and exclusion.

Sub-Saharan Africa has suffered serious adverse affects from globalization. This region now has three million displaced people, fourteen million AIDS orphans, 475 million Africans living on less than the equivalent of $2 per day, while hunger afflicts 40 million people. The devastation resulting from HIV/AIDS in Africa needs to be seen in the context of three hundred years of slavery (1441–1870), seventy-five years of colonialism (1885–1960), and the Cold War (1960s–1991), that successively debilitated the subcontinent. Excitement and pride that came with independence in the 1960s turned to despair under the rule of tyrants in the 1970s. By the 1990s, many viewed African countries as "political and economic infernos."[12] The United States retreat from Africa after the Cold war, in addition to the extraction of resources, including skilled labor, diamonds, and oil, perpetuates centuries of exploitation. Sub-Saharan Africa's debt of $275.6 billion that it can never repay, results in annual interest charges that cancel out the $21.2 billion annual aid donation to Africa, cripples health services, and stultifies development.[13] We need to acknowledge the extent to which the so-called developed world has been instrumental is contributing to such deprivation as described by Thomas Pogge.[14] It must be conceded that corruption and bad government in developing countries contribute significantly to their misery and poor health. Much less openly discussed is the complicity of powerful nations in supporting leaders who are despots and kleptocrats, by legitimizing their right to sell their countries' natural resources, spend profligately on themselves, incurring debts that their impoverished citizens must repay.

4. HIV/AIDS and the Recrudescence of Infectious Diseases

Life is characterized by symbiotic and parasitic relationships between microbes and human and nonhuman animal hosts. Evolutionary changes and the impact of human behavior on the environment have contributed to changes in these relationships and to imbalances favoring sporadic plagues. We saw changes more striking in the second half of the twentieth century than ever before. The world's population has doubled from three to six billion people. Changes in the global economy have resulted in rapidly widening economic disparities with billions of people now living under dehumanizing conditions of poverty and squalor. New patterns of war and ethnic conflict have resulted in displacement and migration of millions of people, and widespread ecological damage has occurred. All of these changes contribute to the development of new ecological niches that alter the balance between microbes and human beings. At the same time, rapid travel and new forms of communication enhance the ability for change anywhere to affect distant others.

Against this background, about thirty new infectious diseases have emerged in the past twenty-seven years. These include Legionnaires disease,

Lyme disease, different forms of encephalitis, bovine spongiform encephalopathy ("Mad Cow" disease), Hantaan virus disease, Ebola fever, and other hemorrhagic diseases, new forms of *E. Coli* and *Vibrio cholera*, Severe Acute Respiratory Syndrome (SARS), and HIV/AIDS, the most threatening of all. Many other infectious diseases such as measles, multi-drug resistant tuberculosis, and malaria, that had become localized, limited, or controlled in the past, have recrudesced with serious adverse implications for all.[15]

The recrudescence of infectious diseases and the HIV/AIDS epidemic have reminded us that people worldwide face common threats, and that infectious diseases may become the greatest security threat to the lives of all people globally. That HIV/AIDS stimulated intense debate on ethical issues—from the micro level of interpersonal relations to the ethics of national relations with transnational pharmaceutical companies, and the ethics of international relations that affect population health—is not surprising.

5. Persisting Controversies in International Clinical Research

Much of the recent controversy on research ethics has arisen from research on HIV/AIDS. For example, the use of placebos in trials of treatments to prevent mother-to-child transmission of HIV excited an intense, ongoing, and often acrimonious debate. In reviewing persisting controversies in international research ethics, Ruth Macklin has described general agreement on the following points:

(1) Research should be responsive to the needs of people in the community we study.
(2) We need research on diseases that occur frequently in poor countries, especially when these cause high morbidity and mortality.
(3) To exploit vulnerable individuals is unethical.
(4) To lower the ethical standards for research in developing countries is unacceptable.[16]

She then lists still unresolved issues within each of these areas of broad agreement. For example:

(1) How do we make decisions about what research to undertake in developing countries, and how do we prioritize these?
(2) What sorts of study designs are acceptable? Can we use placebos, and what comparative arms should be included?
(3) What does it mean not to exploit people?
(4) What is the standard of care? How is this defined and how can it be justified?[17]

Given all these issues, I will limit my discussion to the standard of care that we should provide to research participants.

6. Standard of Care

One of the most hotly debated issues in international collaborative clinical research is the standard of care applied to research subjects in developing countries. The issue first arose in consideration of drugs administered to subjects assigned to control groups, with studies researching prevention of mother-to-child transmission of HIV infection being the most controversial. Critics issued vehement outcries against use of placebos in such studies, based on concern that researchers were not treating subjects in the developing world with the same respect and dignity as subjects in wealthier countries. They contended that this represented a double standard in the ethics of research. Under critical scrutiny, the validity of this argument was found to be too simplistic.[18]

Critics have also argued that a real interest in applying the same standard of care in developing countries as in developed countries would extend beyond drugs to include other fundamental aspects of medical care associated with the research project. These would include availability of appropriate diagnostic technology and provision of adequate nutritional support and treatment for other diseases afflicting research subjects. Follow-up care, monitoring of progress, equivalent expertise of the health-care team, and obtaining informed consent with knowledge of local languages and cultures should be subject to the same standards as in the developed world. While applying all of these standards to all subjects in every country might not be possible, we can conceive the following standards to have universal applicability:

(1) We can conduct research with the same respect for the dignity of all subjects wherever they are in the world, as reflected in treating them as ends in their own right and not using them merely as means to acquire knowledge that could be of benefit to others.
(2) We can ensure that research undertaken in developing countries is in the best interests of subjects in those countries. This means that we place a high priority on diseases most prevalent in the communities from which subjects are recruited, and that we provide care for other diseases afflicting research subjects for which treatment would not otherwise be available.
(3) We can carefully evaluate the long- and short-term risks and benefits of the research, and ensure a fair distribution of risks and benefits.
(4) We can obtain valid informed consent by structuring the process of obtaining consent within the linguistic and cultural framework of research subjects.
(5) We can enhance the potential for community benefit to flow from research by raising the standard of care associated with research to include the components mentioned above.
(6) We can avoid exploitation.[19]

We could achieve a universal standard of care by incorporating these principles into feasible local practices and then ensuring progressive improvements with successive research projects.

The challenge of progressively linking health care to research will engender many complexities that we need to address. For example:

(1) Currently, funding sources often stipulate that the funds are restricted to research activities, limiting what researchers can do with their budget.
(2) Improving standards of care for subjects will result in dual standards of care within poor countries, with research subjects enjoying treatment to which others in the community lack access.
(3) The higher standard of care may become an inducement, but that level of improved care tailored to research may not be sustainable.
(4) Paying higher salaries to researchers privileges them over colleagues who provide regular health services. Such studies can add to the brain drain of qualified staff away from the public health sector.
(5) The availability of additional care contributes to subjects' misunderstanding that the trial is about treatment and not research.[20]

We can offer some rational solutions to ameliorate these dilemmas. For example, while limitations on how researchers can use research funds are a reality, researchers could include a small budget for associated expenses in their grant applications. Doing so could progressively sensitize granting agencies to recognize the validity of such components of the budget. In addition, researchers could seek out other aid agencies from their countries who are working in the same geographic region to develop partnerships that could be of value to the research and to the aid endeavor. National research councils in wealthy nations could facilitate such coordination for their researcher through central databases of aid projects.

Dual standards of care ensuing from the above activities are arguably an inevitable aspect of progress. Inability to achieve immediate equity should not be an impediment to making improvements that could spread more widely with time and effort. We can argue that to be satisfied with existing low or nonexistent standards is less ethical than to create dual standards as part of a progressive means of improving care. True, a higher standard of care will be an inducement. But inducements are only morally wrong if they result in participants taking risks with their lives. What is wrong with an inducement that enables access to otherwise unavailable care? Inability to sustain high salaries for researchers is a problem, but this, too, could be used as leverage towards better conditions for local health workers. Finally, we can overcome misconceptions about the role of research through good communication.

We can give many reasons why the overall standard of care for research subjects is critical. These include moral arguments that embrace the concern not to do harm, the desire to do good, and the aim to be fair. Other moral goals include having respect for other cultures and sensitivity to the invasive impact of itinerant researchers from developed countries. Strategically, we can argue that an improved standard of care could enhance participation in research and the achievement of research goals. In addition, we can mount an operational argument in support of the idea that research coupled with improved standards of care facilitates improvements in the delivery of health care in general. As the goal of medical research is to improve health care, there should be greater support for encouraging closer links between research and its application at sites where the research is being undertaken.[21]

Informed consent is one of the major underlying principles of research. If we believe all that we say and claim about informed consent, we would ensure that researchers of the same culture and language group as the research subjects obtain the consent. In reality, having highly trained people obtaining informed consent is rare. A multi-page translation of an informed consent document is not the solution. A legitimate informed consent decision-making process requires the ability to overcome cultural and linguistic barriers. We tend to focus on scientific skills, equipment, and methods in research, whereas communication skills and trust are equally essential and require allocation of adequate resources. Currently the standard for obtaining ethically legitimate informed consent is under-emphasized at the practical level.

7. Making Progress through New Perspectives

Instead of undertaking research autocratically on passive subjects in poor and developing countries, we should form partnerships with developing country co-investigators and communities who should be actively involved in the design and conduct of trials. In this way, privileged researchers can build the capacity of less privileged colleagues while they build their capacity through better understanding of the communities and participants with whom they are interacting.

We need to pay greater attention to the potential harms of research. These extend beyond the physiological dangers that have been the traditional focus of concern. We also need to consider psychosocial risks for individuals and for communities.

Potential benefits go beyond benefits to individuals. They include benefits to communities and sponsors such as academic institutions and pharmaceutical companies. If we have a genuine interest in justice and fairness, then we must examine much more closely than is currently the case whether benefits to the sponsors balance benefits for participants and communities.

By linking research to improved health care, we can provide an impetus to operationalize research. Such progress requires acknowledgment that finding a reasonable middle path between pragmatic unprincipled approaches and ideological unachievable utopias is possible. We should also develop partnerships with other aid organizations to facilitate the achievement of mutual development goals.[22]

All of this requires a new paradigm of thinking that includes greater concern about population health, careful balancing of individual and social goods, and attention to the widening economic disparities that threaten our humanity. Progress will depend on recognizing that the ethics of international research goes beyond micro-ethics. While the ethics of interpersonal relationships are critical, we also have to think ethically at a wider level. We need to expand the discourse on ethics and human rights beyond interpersonal ethics and civil and political rights based on the ideas of individual rights and freedom. We should pay greater attention to the ethics of how institutions operate and interact, to the ethics of public health, to social and economic rights based on concern for equity, and to the ethics of international relations that affect whole populations, based on solidarity.[23]

Finally, we should endeavor to link research to development by going beyond a paternalistic approach to development towards an emancipatory framework, one that uplifts people from dependence and facilitates independence. A form of participatory evaluation, one in which people participate in decision making through dialogue and debate instead of having decisions imposed on them, is also required to confirm that we are helping people from their perspective, and not merely from ours.[24] Capacity must be built through interdisciplinary work, with greater understanding of how complex systems function and with application of systems theory to improving health care.

8. Conclusion

The ethics of research is changing in response to a changing world, but I believe that change is not rapid enough. Neither the interests of pharmaceutical companies nor scientists should dominate research agendas. We should also respect the justifiable moral aspirations of individuals and their communities to improve their health. If we can shift our mindset to these new directions, then we will be able to keep pace with a rapidly changing world that demands adaptation. Increased short-term costs will pay the dividend of increased long-term benefits.

NOTES

1. Anonymous, "Safeguarding Participants in Clinical Trials," *Lancet*, 355 (2000), p. 2177.

2. Ruth L. Sivard, *World Military and Social Expenditures 1996*, 16th ed. (Washington D C: World Priorities Press, 1996).

3. Commission on Health Research for Development, *Health Research: Essential Link to Equity in Development* (Oxford: Oxford University Press, 1990).

4. Pierre Trouiller, P. Olliaro, Els Torreele, James Orbinski, Richard Laing, and Nathan Ford, "Drug Development for Neglected Diseases: A Deficient Market and a Public Health Failure," *Lancet*, 359 (2002), pp. 2188–2194.

5. Robert Black, Saul S. Morris, and Jennifer Bryce, "Where and Why Are 10 Million Children Dying Every Year?" *Lancet*, 361 (2003), pp. 2226–2234.

6. Gillian Walt, "Globalization and Health," *Medicine, Conflict and Survival*, 17 (2001), pp 63–70.

7. Richard J. Barnet and John Cavanagh, *Global Dreams: Imperial Corporations and the New World Order* (New York: Simon and Schuster, 1994).

8. Sivard, *World Military and Social Expenditures 1996.*

9. Ann Pettifor, *Debt: The Most Potent Form of Slavery* (London: Christian Aid Society, 1996); and Ann Pettifor and Janet Bush, eds., *Real World Economic Outlook: The Legacy of Globalization: Debt and Deflation* (New York: Palgrave Macmillan, 2003).

10. Wayne Ellwood. *No-Nonsense Guide to Globalization* (London: Verso, 2003).

11. Barry Gills and Joel Rocamora, "Low-Intensity Democracy," *Third World Quarterly*, 13:3 (1992), pp 501–523.

12. Peter Schwab, *Africa: A Continent Self-Destructs* (New York: Palgrave Macmillan, 2002).

13. "Into Africa: Special Report," *Toronto Star*, 25 May 2003.

14. Thomas Pogge, *World Poverty and Human Rights* (Cambridge, UK: Polity Press, 2002).

15. *World Health Report 1996* (Geneva: World Health Organization, 1996); and Laurie Garrett, *The Collapse of Public Health* (New York: Hyperion, 2000).

16. Ruth Macklin, "After Helsinki: Unresolved Issues in International Research," *Kennedy Institute of Ethics Journal*, 11:1 (2001), pp. 17–36.

17. Ibid.

18. Alex J London, "The Ambiguity and the Exigency: Clarifying 'Standard of Care' Arguments in International Research," *Journal of Medicine and Philosophy*, 25:4 (2000), pp. 379–397; and Alex J. London, "Equipoise and International Human-Subjects Research," *Bioethics*, 15:4 (2001), pp. 312–332.

19. Solomon R. Benatar and Peter A Singer, eds., "A New Look at International Research Ethics," *British Medical Journal*, 321(2000), pp. 824–826.

20. K. Shapiro and Solomon R. Benatar, "HIV Prevention Research and Global Inequality: Towards Improved Standards of Care," *Journal of Medical Ethics*, 31 (2005), pp. 39–47.

21. Ibid.

22. Bernie Lo and Ronald Bayer, "Establishing Ethical Trials for Treatment and Prevention of AIDS in Developing Countries," *British Medical Journal*, 327 (2003), pp. 337–339.

23. Solomon R Benatar, Abdallah S. Daar, and Peter A. Singer, "Global Health Ethics: The Rationale for Mutual Caring," *International Affairs*, 79 (2003), pp. 107–138.

24. Ronald J. Kushner, T. G. Baker, "Never the Twain Will Meet: Cross-Cultural Evaluation of Organizational Performance," Presented to the Association for Research on Non-Profit Organizations and Voluntary Action, Seattle, Washington, 7 November 1998.

Two

SOME CURRENT ISSUES IN THE ETHICS OF BIOMEDICAL RESEARCH AND THEIR BACKGROUND IN THE PROTECTION OF THE DIGNITY AND AUTONOMY OF THE VULNERABLE

Matti Häyry

1. Themes and Backgrounds

In this chapter, I will identify some contemporary themes in the ethics of biomedical research and sketch the bioethical background against which we can best understand them. Four issues that prevail in current debates are:

(1) The moral questions of freedom of research and sanctity of life in human embryonic stem cell research;
(2) The political rhetoric of risk and precaution in research into genetically modified organisms;
(3) The interplay of solidarity and justice in global and gender issues concerning the development and marketing of pharmaceuticals; and
(4) The paradox of ethical regulation and trust in public perceptions on bioscientific advances.

Before I address these topics, I will set the scene by examining in some detail two underlying issues:

(1) How is bioethics practiced? What are its potential scope and approaches?
(2) Can we find some unity in bioethics despite all the diversity in the field?

Answers to these questions should pave the way to a better understanding of the current debates.

2. Bioethics: Its Scope and Approaches

Four partly overlapping areas fall within the scope of bioethics. These are:

(1) The professional codes of people working in medicine, health care, and related fields;
(2) The social and political organization of health-care delivery by professionals, governments, and global and local authorities;
(3) The arrangement and regulation of scientific and academic research into medicine, health care, biology, genetics, and related fields; and
(4) The proper and fair management of living resources, including the exploitation and preservation of our natural environment.

We used to call the first area medical ethics or nursing ethics, but the term professional ethics is now gaining ground as a more general name.[1] Ethical work in this field has also been called health care ethics and bioethics.[2]

Research ethics is perhaps the most common name for the third activity, although the meaning of the word bioethics has, in some circles, been applied solely to this area.[3] As for the fourth area, this was arguably the original denotation of bioethics, although environmental ethics may now be a more popular term, and global ethics and development ethics appear to claim parts of this and other areas of bioethics.[4] Biomedical ethics is also a name sometimes applied to bioethical activities.[5]

The approaches taken to the practice and study of bioethics are equally diverse. They include professional self-regulation, legal control, moral sociology, history, gender studies, cultural anthropology, philosophy, theology, and the contributions made by different interest groups.

Originally, only physicians practiced professional self-regulation (from the establishment of the Hippocratic Oath on). Then nurses, social workers, engineers, scientists, and professional groups adopted the practice.[6] When professionals argue that they do not need ethics components treated separately in their curricula, they usually claim that good practice and right conduct result from good professional training and self-regulation, not from knowing what ancient philosophers have said about morality.

Sometimes called medical law or biolaw, the legal approach extends jurisprudential thinking and legislative action into all areas of bioethics.[7] To the surprise and dismay of many academic ethicists, practical legal regulation has become quite a popular form of bioethics, with presidents and parliaments often implementing quick bans on activities to please their constituencies.

Academics working in the social sciences and humanities conduct bioethical studies in moral sociology, history, gender studies, and cultural anthropology.[8] Many medical schools have preferred this approach to philosophical and theological ethics because instead of giving ready-made intellectual solutions, people in sociology, history, gender studies, and literature can arguably empower students by prompting them to think about their future work in different ways.

Philosophical and theological approaches could obviously be included in the humanities, but they are worth mentioning separately, since they are arguably the disciplines that started the modern bioethics movement in the West half a century ago. Philosophers in the field see it as their job to increase conceptual clarity. Similarly, I would imagine that theologians participate in the discussion to raise awareness concerning widely shared religious traditions. In any case, philosophers and theologians do not always see themselves as advocates of particular positions or views, although others often appear to think that they are.[9]

Professions, political parties, and religious groups have participated in bioethical debate since its beginning in the 1950s and the 1960s. Later, patient associations, consumer groups, and many other practical and ideological groupings in the field emerged and came to be seen as the operative interest groups in the field.

In addition to these four general scope categories, which can be mixed, and five main approaches, which can be mixed and combined, other diversifying considerations can be identified. Different people hold different views on the relationship of theory and practice in bioethics. Some believe that bioethics is about understanding some of these things, while others think that the point is not to understand or explain the world, but to change it for the better. The emphasis in people's bioethical work can vary considerably according to their choice in this matter.

3. Dignity and Autonomy of the Vulnerable

So the question is, can any unity be found amidst all this diversity of scopes and approaches? A sensible person would state unambiguously, "No." Any attempt to capture the variety of bioethics in only one view or definition is bound to be one-sided, partial, and, as a description of the whole field, plainly mistaken.

My answer to the same question is, "Perhaps." I will put forward a notion that might determine the essence of most activities that we call "bioethical." I offer this notion not as the absolute truth, but as a hypothesis that can be tested by applying it to different practices and developments, and then checking how well or how poorly it conforms to them. So here goes.

My thesis, for the purposes of the rest of this chapter, is that bioethics, as an activity and as a discipline, is concerned with the dignity and autonomy of vulnerable individuals and populations, especially in medicine, health care, medical, pharmaceutical, and biological research, and environmental issues, and possibly in other related fields, as well. I submit that everyone can accept this general description, but I also believe that people understand it in different ways. Disputes arise when we start to debate the meaning of central terms: "What is dignity?" "What is autonomy?" "Who are the vulnerable?" "Who or what is a threat to the vulnerable?" "What can or should be done to protect the vulnerable, and by whom should it be done?"[10]

4. The First Phase of Bioethics

Let me illustrate the point concerning the uncontroversial nature of my characterization by outlining what I would like to call the first phase of the bioethics movement in the West.

In the 1960s, Western philosophers and theologians started to question some of the then current policies on abortion, euthanasia, and resource allocation in medicine and health care.[11] Some thought, for instance, that the abortion policies they criticized were too restrictive, and did not sufficiently respect the dignity and autonomy of women. Others argued that restrictive policies were justified, because they protected the dignity and autonomy of unborn human beings—fetuses.

In a similar vein, some ethicists believed that keeping patients alive against their considered wishes is a violation of human dignity and autonomy. Others thought that since we cannot autonomously choose to die, passively letting them die, or killing them at their request, would be an act against dignity and autonomy. In this dispute, both sides claimed that they were concerned about the dignity and autonomy of the vulnerable. The differences came from different readings of dignity, autonomy, and their protection, and, most of all, from disagreements about who are the most vulnerable. Should we grant this status to the women denied abortions or the aborted fetuses? Should we give it to the suffering patients who are denied assistance to die or to the patients whose death is directly or indirectly hastened?

A possible objection to my interpretation is that the disagreements are too strong to allow any common goals, or common agenda. Labels like "liberal" and "conservative" have been attached to the warring parties of the abortion and euthanasia debates, and some analysts have opined that no consensus can possibly be found between these groups.

But people do not fall into categories quite that easily. To illustrate my point, consider the third early issue in bioethics, the allocation of scarce medical resources. People who had been advocating the freedom of choice, often "liberals" with some "socialist" tendencies, suddenly found themselves arguing from the same premises as the champions of the free market, who would have been happy to close down all public hospitals and health centers. We find similar inconsistencies in the other camp. Some people who had been "conservative" in the abortion and euthanasia debates, notably on some Christian grounds, now defended public health programs with some of the "liberals."

The idea of protecting the vulnerable ran through all aspects of the early bioethics movement. Despite their differences of opinion, most bioethicists during this first phase recognized a common enemy, namely the medical profession and the paternalistic attitudes of its members toward their patients and clients and the rest of society. Abortions were sometimes permitted and sometimes denied, but the decisions were, in most cases, made by physicians. The same observation applies to the euthanasia issue. People were sometimes allowed to die and sometimes kept alive against their wishes, again, by health-care professionals. The allocation of resources was sliding into the hands of medical professionals, despite the more general tendencies to make decisions like this more and more democratically.

The common denominator of the first phase of the bioethics movement was (and to a degree still is—these issues have not gone away) an opposition to the paternalistic power of physicians and, to a lesser extent, other health-care professionals, and the protection of the dignity and autonomy of ordinary citizens, who were seen as the vulnerable group in the jaws of the medical machine.[12] This common ground, I contend, made possible the creation of a relatively unified discipline out of these—sometimes quite fierce—quarrels between pro-life and pro-choice, radical and conservative, liberal and socialist academics. They may not have noticed, but the pioneers of bioethics were, in most cases, united by the conviction that medical or scientific expertise does not give professionals any legitimate authority in moral matters.

5. The Second Phase of Bioethics

Since the 1990s, and in some areas earlier, we have been entering the second phase of the bioethics movement, characterized by a tendency opposite from that found in the first phase, namely a tendency to empower professionals in ethical matters and to put them back in charge of their professional decisions.[13]

Bioethics teaching in medical schools has, from the beginning, included this aspect of empowerment.[14] One of the aims of bioethics curricula has been to turn physicians and nurses into ethically competent decision makers and good professionals in their respective fields of study. The question is whether we can achieve this goal by exclusively training students in good professional practice, as understood by their clinical teachers, or whether ethics should enter the curriculum in some other sense. The emergence of teachers—mentors who can arguably incorporate ethical theories and external concerns into their professional teaching—with double qualifications in ethics and other fields has partly answered this question.

The shift from the battle against paternalism to the empowerment of the professionals is visible in recent discussions that focus on the duties of patients.[15] During the first phase of bioethics, few academics had the audacity or courage to suggest that recipients of health-care services should assume a responsible role in their treatment. Patients were seen as victims of social arrangements and medical paternalism, and obligations were assigned to professionals, not to their clients. More recently, saying that patients, too, have health-related duties, and that they should share the responsibility for the success of treatments with their health-care providers, has become more acceptable.

In medical, biological, and genetic research, we can see the empowerment trend in two interrelated developments. The first is that we have left the regulation of sensitive matters, in many fields, primarily to the self-assessment of people involved in research. To ensure the quality of this self-regulation, governments have increased their control over institutional ethics committees, and in some cases required that institutional ethics committees be established, but they have, at the same time, left the screening of research activities and the responsibility for the criteria of assessment, to the members of the committees.[16]

The second development is that in some regions of the world, the public is now being educated to realize the benefits of scientific advances. In the United Kingdom, for instance, this education is, at the turn of the millennium, partly provided by governmentally funded "genetic knowledge parks," one point of which is to increase the public understanding of science. This education contributes to the smooth running of ethics committees because the obligatory lay members of these committees (needed in the name of openness, widening participation, and democracy), through the education process, will become more lenient towards forms of research that would otherwise have raised—possibly unnecessary—fears in them.

The—possibly controversial—justification for the empowerment of medical professionals and scientists can, again, be found in considerations of the autonomy and dignity of the vulnerable. If we can help people by medicine, genetics, and the like, then we can better protect the self-determination and worth of vulnerable individuals by eliminating fears than by banning useful activities in an attempt to accommodate everybody's feelings. Those who do not buy into this kind of thinking, in their turn, can argue that the anxieties are real, and that we can better protect the truly vulnerable by controlling scientific developments than by trying to manipulate people into approving them.

6. Dignity, Autonomy, and Stem Cell Research

I turn my attention now to some issues regarding stem cell research, genetically modified food products, international drug development and marketing, and the relation between regulation and trust as they pertain to my thesis that bioethics, as an activity and as a discipline, is concerned with the dignity and autonomy of vulnerable individuals and populations.

The scientific background of the stem cell debate is as follows: Stem cells are undifferentiated cells that can either divide and multiply in their original state, or become differentiated into other, more specialized types of tissue. Some of them are totipotent, which means that they are capable of developing into a complete organism, or differentiating into any of its cells and tissues. Others are pluripotent, which entails that while their range of powers is limited, they can still produce several kinds of tissue. Stem cells are found in many parts of the adult human body, where their function is to replace tissue lost through ageing or trauma. The inner cells of early embryos also belong to this category. During fetal growth, stem cells evolve into all the different tissues.[17]

Some therapies, notably bone marrow and skin transplants, have been based for some time on knowledge of these facts. But until recently, we had no method of keeping generations of stem cells alive in their totipotent or pluripotent state. In 1998, scientists at the University of Wisconsin published a pioneering method by which human embryonic stem cells can be cultured indefinitely, which, in turn, makes possible the creation of stable cell lines for research and the development of therapies.[18]

Human stem cells can be produced in three distinct ways. The first is nuclear transfer, which has also been used in the reproductive cloning of mammals since 1997, when Dolly, the cloned sheep, was created by researchers in Scotland.[19] In this method, the original nucleus of an embryo is removed and replaced by the nucleus of another cell from the same or a different individual. The second method is to harvest human stem cells from early embryos destroyed in the process. Some people make a distinction between embryos intentionally created for research and embryos that are the by-product of reproductive technologies, but any deep significance of this distinction is uncertain. The third way is to isolate adult stem cells and culture them. Most commentators view the third method as uncontroversial.

In many diseases, the human body has lost, partly or entirely, its ability to produce some vital cells or tissue. Stem cell therapies can, if suitable cells are produced in sufficient quantities, reintroduce these to the system, and thereby cure the diseases or alleviate their symptoms. In the future, it may also become possible to create entire organs in the laboratory for transplantation. This would make organ donations unnecessary and ensure that the tissue used in transplants is perfectly compatible. In addition, some rejuvenating treatments could be developed that might significantly extend the human life span.[20]

The bioethical debate concerning stem cells has mostly centered on the destruction of human embryos in research and in therapies. Some argue that destroying embryos violates their dignity and autonomy, while others say that embryos do not yet qualify as human beings who have these ethical qualities. This polarization has led many ethicists to believe that the stem cell discussion is a mere repetition of the abortion debate of the 1960s and the 1970s, since in both, freedom and sanctity are set against each other. While this view is partly correct, current bioethical disputes have new features that we need to take into account.

According to my hypotheses, everyone is interested in the dignity and autonomy of the vulnerable (although these can be defined in different ways), and most bioethicists are now anxious to empower scientists in their work. We can easily see the empowerment aspect in the permissive view, which aims to protect the freedom of research in all areas, with the ultimate goal of promoting the health of humankind (which is vulnerable to diseases). We can detect—perhaps surprisingly—the same logic in attempts to ban human embryonic stem cell research, human reproductive cloning, and other activities that may alarm the public. By forbidding some forms of research, lawgivers in many countries try to ensure that other, related, forms of research can continue without interruption. The authorities hope that once public trust has been secured by some prompt legislative action, the citizens rest assured that the forms of research that are not prohibited are seen as ethically acceptable.[21]

Concentration on the sanctity of the embryo dispute tends to overshadow other ethical questions in stem cell research. Critics with more moderate opinions have pointed out that the resources allocated to this type of science tend to drain the resources of other kinds of research, which could be more helpful to humanity in the end. These critics have noted that twenty years ago people were hopeful that gene therapies would be curing many diseases by now; yet next to nothing has been achieved, in terms of standard therapies, in this field. In addition, while some future gene therapies could be successful in affluent societies, the likelihood that people in the third world would experience any of the benefits in the near future is extremely low. Again, this is a disagreement about who are the most vulnerable and how we can most efficiently protect their dignity and autonomy.

7. Political Rhetoric and Genetically Altered Food Products

Slightly different aspects of current bioethical debates come into focus in the context of the other three topics listed. The current discussion on the advantages and disadvantages of genetically modified organisms in agriculture and food production provides a nice example of the use of bioethical rhetoric in political disputes.

In Europe, people appear to be reserved when it comes to genetically modified products. The idea behind this is precaution—we have no knowledge of any health risks caused by, say, American tomatoes, but we feel that they might be dangerous.[22] On the strength of this feeling, the European Union has succeeded in keeping modified foods marketed by American companies out of the European continent, a policy that could be explained by economic self-interest besides ethical sensitivity.

In the United States, on the other hand, the genetic modification of food products is well advanced, and the majority of people do not appear to have too many problems with it. On the strength of this, and in the name of national interest, United States' authorities have tried to persuade other countries to lift their trade restrictions. President George W. Bush, who declared that European politicians, by their attitude, are responsible for famine in the third world, launched one strike in this battle. The European authorities in question were less than enthusiastic to hear this accusation.

Although I cannot immediately see the connection with the empowerment of professionals, unless we are talking about professional food producers, the protection of the autonomy and dignity of the vulnerable looms large on both sides of the debate. European leaders argue that they are protecting their vulnerable citizens and food producers; whereas President Bush argued he was promoting the dignity and autonomy of the vulnerable populations of the poorer countries.

8. Justice, Solidarity, and International Pharmaceutical Research

The development and international marketing of pharmaceuticals raises many ethical concerns, which are the topic of other chapters in this book. One of these is research in third-world countries where regulations and restrictions may be fewer, but where the participants of the trials may be left without proper medication after the completion of the research.[23] Another problem is posed by patent and price policies which may provide an incentive for the costly development of new drugs, but which may also create economic difficulties for effective health-care provision in poorer regions of the world. Another dilemma is created by prohibitions on testing drugs on children and pregnant women. This may protect them from the side effects of research, but it also leads to situations where drugs not tested on matching populations are prescribed to them, or they are left without proper medication.

All these issues are debatable in terms of justice and solidarity among individuals, communities, and nations. Often the debates in this relatively new field of bioethics are strongly polarized and resemble in intensity the earlier abortion and euthanasia disputes.

Quite possibly the discussion on some of these issues could benefit from attempts to empower the professionals in the field. Researchers and pharmaceutical companies should always disclose any genuine conflicts of interest in the development and marketing of drugs, and we should always keep the professional power of researchers in check. But we must also realize that some drugs do save lives and improve health, and that they can promote the autonomy and dignity of humankind as much as the protection of special groups against the hardships of research.

9. The Paradox of Trust

I have suggested in this chapter that one of the legitimate aims of bioethical regulation is to make the public more convinced that what doctors, nurses, health administrators, scientists, and environmental authorities do is good and right, and that these professionals do the best they can under the circumstances. Recently critics have argued that bioethical regulation fails to achieve this goal. The phenomenon to which is alluded in this argument can be called "the paradox of trust." Its logic follows.[24]

People have fears and anxieties concerning scientific advances and their application to social life—an example is the establishment of national genetic data banks. Authorities respond to these fears and anxieties by putting in place control mechanisms: ethics committees, audit systems, and the like. These mechanisms make the institutions applying the results of research more trustworthy, objectively speaking. Yet this is not what members of the public perceive. They see that now another layer of bureaucracy faces them, which makes it even harder for them to see whether they can trust the people involved. Besides, when they see all the control mechanisms, they quite naturally ask, "Why would these be needed if we could trust them in the first place?" The mechanisms that make institutions more trustworthy make people trust them less. This is the paradox.

Is there any way out of this situation? Well, some bioethicists say that we should nurture an environment of trust by encouraging collective, communicative ways of thinking and doing things instead of treating people as separate individuals with only their personal goals in mind.[26] But how can we achieve this goal? If the answer is, "By putting in place mechanisms which would promote this end," we add another layer of bureaucracy, and defeat the purpose of what we aimed to do.

10. Construction by Destruction

In conclusion, my tentative recommendations in this chapter have gone in two opposing directions. Where I have seen sharp disagreements, I have suggested that it would be beneficial to find a common language. Where I have seen consensus, I have suggested that we should tear apart this consensus to find new grounds for disagreement.

This reflects my view on what philosophers can do in bioethical debates. If people are too far apart to speak to each other, they need to come together so that they can enter a meaningful dialogue. But if the opposing parties to a debate appear to sit too cozily together, they should be reminded that they may just have forgotten in what the people they represent believe.

ACKNOWLEDGMENTS

This article was produced as a part of *Ethical, Legal and Social Aspects of Human Genetic Databases: A European Comparison* (ELSAGEN), financed between 2002–2004 by the European Community (QLG6-CT-2001-00062), and *Ethical and Social Aspects of Bioinformatics* (ESABI), financed between 2004–2007 by the Academy of Finland (SA 105139). My thanks are due to Peter Herissone-Kelly for checking my English.

NOTES

1. Albert R. Jonsen, *A Short History of Medical Ethics* (New York: Oxford University Press, 1999); Elsie L. Bandman and Bertram Bandman, *Nursing Ethics through*

the Life Span, 4th ed. (Upper Saddle River, N.J.: Prentice Hall, 2002); and Ruth Chadwick, ed., *Ethics and the Professions* (Aldershot, England: Avebury, 1994).

2. John F. Monagle and David C. Thomasma, *Health Care Ethics: Critical Issues for the 21st Century*, 2nd ed. (Boston: Jones & Bartlett Publishers, Inc., 1997).

3. Kåre Berg and Knut Erik Tranøy, eds., *Research Ethics* (New York: Alan R. Liss, 1983); and Darwin Cheney, ed., *Ethical Issues in Research* (Frederick, Md.: University Publishing Group, 1993).

4. Van Rensselaer Potter, *Bioethics: Bridge to the Future* (Englewood Cliffs, N.J.: Prentice-Hall, 1971); Louis P. Pojman, ed., *Environmental Ethics: Readings in Theory and Application* (Belmont, Calif.: Wadsworth, 2001); Martin Shaw, ed., *Politics and Globalisation: Knowledge, Ethics, and Agency* (London: Routledge, 1999); Martha Craven Nussbaum and Jonathan Glover, *Women, Culture, and Development* (Oxford: Clarendon Press, 1995); Kenneth Aman, ed., *Ethical Principles for Development: Needs, Capacities, or Rights?* (Upper Montclair, N.J.: Institute for Critical Thinking, 1991); and Robin Attfield and Barry Wilkins, eds., *International Justice and the Third World* (London: Routledge, 1992).

5. Tom L. Beauchamp and James F. Childress, *Principles of Biomedical Ethics*, 5th ed. (New York: Oxford University Press, 2001).

6. Albert R. Jonsen, Mark Siegler, and William J. Winslade, *Clinical Ethics: A Practical Approach to Ethical Decisions in Clinical Medicine*, 5th ed. (New York: McGraw Hill, 2002); Anne H. Bishop and John R. Scudder, Jr., *Nursing Ethics: Holistic Caring Practice*, 2nd ed. (Boston: Jones and Bartlett, 2001); Francis L. Macrina, *Scientific Integrity: An Introductory Text with Cases*, 2nd ed. (Washington, D.C.: ASM Press, 2000); and Matti Häyry, "Do Bioscientists Need Professional Ethics?" Matti Häyry and Tuija Takala, eds., *Scratching the Surface of Bioethics* (Amsterdam and New York: Rodopi, 2003), pp. 91–97.

7. J. K. Mason and R. A. McCall Smith, *Law and Medical Ethics*, 6th ed. (London: Butterworths, 2002); and John Tingle and Alan Cribb, eds., *Nursing Law and Ethics*, 2nd ed. (Oxford : Blackwell Science, 2002).

8. Rita Charon and Martha Montello, *Stories Matter: The Role of Narrative in Medical Ethics* (London and New York: Routledge, 2002); Ronald A. Carson and Chester R. Burns, eds., *Philosophy of Medicine and Bioethics: A Twenty-Year Retrospective and Critical Appraisal* (Dordrecht: Kluwer Academic Publishers, 1997); Raymond DeVries and Janardan Subedi DeVries, eds., *Bioethics and Society: Constructing the Ethical Enterprise* (Englewood Cliffs, N.J.: Prentice Hall, 1998); Anne Donchin and Laura Purdy, eds., *Embodying Bioethics: Recent Feminist Advances* (Lanham, Md.: Rowman & Littlefield, 1998); Rosemarie Tong, *Feminist Approaches to Bioethics: Theoretical Reflections and Practical Applications* (Boulder, Colo.: Westview Press, 1997); and Susan Wolf, ed., *Feminism and Bioethics: Beyond Reproduction* (New York: Oxford University Press, 1996).

9. Peter A. Singer, *Practical Ethics*, 2nd ed. (Cambridge, England: Cambridge University Press, 1993); Raanan Gillon, *Philosophical Medical Ethics* (Chichester: John Wiley & Sons, 1985); John Harris, *The Value of Life: An Introduction to Medical Ethics* (London: Routledge, 1985); Raanan Gillon, ed., *Principles of Health Care Ethics* (Chichester: John Wiley and Sons, 1994); Stephen E. Lammers and Allen Verhey, eds., *On Moral Medicine: Theological Perspectives in Medical Ethics*, 2nd ed. (Grand Rapids, Mich.: Eerdmans, 1998); H. Tristram Engelhardt, Jr., *The Foundations of Bioethics*, 2nd ed. (New York: Oxford University Press, 1996); H. Tristram Engelhardt, Jr., *The Foundations of Christian Bioethics* (Lisse: Swets & Zeitlinger Publishers, 2000);

and B. Andrew Lustig, ed., *Bioethics Yearbook Volume 5—Theological Developments in Bioethics: 1992–1994* (Dordrecht: Kluwer Academic Publishers, 1997).

10. Matti Häyry, "Another Look at Dignity," *Cambridge Quarterly of Health Care Ethics*, 13:1 (Winter 2004), pp. 7–14; and Matti Häyry, "Prescribing Cannabis: Freedom, Autonomy, and Values," *Journal of Medical Ethics*, 30:4 (2004): pp. 333–336.

11. Gregory Pence, ed., *Classic Works in Medical Ethics: Core Philosophical Readings* (New York: McGraw Hill, 1998).

12. Donald Van DeVeer, *Paternalistic Intervention: The Moral Bounds on Benevolence* (Princeton, N.J.: Princeton University Press, 1986); Gerald Dworkin, *The Theory and Practice of Autonomy* (Cambridge, England: Cambridge University Press, 1988); and Heta Häyry, *The Limits of Medical Paternalism* (London and New York: Routledge, 1991); cf. Ivan Illich, *Limits to Medicine—Medical Nemesis: the Expropriation of Health* (London: Marion Boyars Publishers Ltd., 1976).

13. Tom Koch, "Absent Virtues: The Poacher Becomes Gamekeeper," *Journal of Medical Ethics*, 29:6 (2003), pp. 337–342.

14. Terry R. Bard, *Medical Ethics in Practice* (New York: Hemisphere Publishing Corporation, 1990).

15. John Harris and Søren Holm, "Is there a Moral Obligation not to Infect Others?" *British Medical Journal*, 311:7014 (1995), pp. 1215–1217; and Rosamond Rhodes, "Genetic Links, Family Ties, and Social Bonds: Rights and Responsibilities in the Face of Genetic Knowledge," *Journal of Medicine and Philosophy*, 23:1 (1998), pp. 10–30; cf. Heather Draper and Tom Sorell, "Patients' Responsibilities in Medical Ethics," *Bioethics*, 16:4 (2002), pp. 335–352.

16. The European Group on Ethics in Science and New Technologies to the European Commission, *Ethically Speaking* 2 (Luxembourg: Office to Official Publications of the European Communities, 2003).

17. Søren Holm, "Going to the Roots of the Stem Cell Controversy," *Bioethics*, 16:6 (2002), pp. 493–507.

18. J. A. Thomson, J. Itskovitz-Eldor, S. S. Shapiro, M. A. Waknitz, J. J. Swiergiel, V. S. Marwill, and J. M. Jones, "Embryonic Stem Cell Lines Derived from Human Blastocysts," *Science*, 282:5391 (1998), pp. 1145–1147.

19. I. Wilmut, A. E. Schnieke, J. McWhir, A. J. Kind, and K. H. S. Campbell, "Viable Offspring Derived from Fetal and Adult Mammalian Cells," *Nature*, 385:6619 (27 February 1997), pp. 810–813.

20. John Harris, *Clones, Genes, and Immortality: Ethics and the Genetic Revolution* (Oxford: Oxford University Press, 1998); and Shirley Wright, "Human Embryonic Stem cell Research: Science and Ethics," *American Scientist*, 87:4 (1999), pp. 352–361.

21. Tuija Takala and Matti Häyry, "Benefiting from Past Wrongdoing, Human Embryonic Stem Cell Lines, and the Fragility of the German Legal Position," *Bioethics* (forthcoming, 2006).

22. Matti Häyry, "European Values in Bioethics: Why, What, and How to Be Used?" *Theoretical Medicine and Bioethics*, 24:3 (2003), pp. 199–214.

23. *The Ethics of Research Related to Health Care in Developing Countries* (London: Nuffield Council on Bioethics, 2002).

24. Cf. Onora O'Neill, *Autonomy and Trust in Bioethics: The Gifford Lectures 2001* (Cambridge, England: Cambridge University Press, 2002); and Peter Herissone-Kelly, "Bioethics in the UK: Genetic Screening, Disability Rights, and the Erosion of Trust," *Cambridge Quarterly of Health Care Ethics*, 12:3 (Summer 2003), pp. 235–241.

25. Michael Parker, "Genetics and the Interpersonal Elaboration of Ethics," *Theoretical Medicine and Bioethics*, 22:5 (2001), pp. 451–459.

Three

BIOETHICS AND BIOMEDICINE: DEVELOPING COUNTRIES' PERSPECTIVE

Florencia Luna

1. Introduction

This paper explores present and possible relations between biomedicine, and the current states of affairs and pressing needs of developing countries. I will examine advantages and failures of sophisticated technologies within the socioeconomic context of non-industrialized countries.

For the purpose of this paper, I will use the term "traditional medicine" to mean that based pharmaceutics and mechanics. In contrast, by "biomedicine," I mean relatively new therapies, treatments, or diagnoses based on recombinant DNA and cellular biology. I also include the so-called reprogenetics—new assisted reproductive technologies such as *in vitro* intracytoplasmatic sperm injection (*in vitro*-ICSI) in addition to recombinant DNA techniques.

After discussing some general empirical information on the health status and socioeconomic situation of developing countries, I will consider the implementation and adoption of existing techniques in biomedicine and some of the problems it faces, such as the challenges to and controversies regarding genetics and assisted reproduction in developing countries. Finally, I will examine the positive and negative potentialities of these new lines of research and examine how we should redirect them towards the needs of developing countries.

2. Causes of Illness and Mortality in Developing Countries

Approximately 80 percent of the world's population resides in countries with gross national products significantly lower than that of highly industrialized nations. The term usually applied to these countries is "developing," but several indicators suggest that the economic situation in the majority of those countries is stagnant or even deteriorating.[1] No single indicator explains the economic, cultural, social, and political complexities of underdevelopment. So developing countries is a controversial category. These countries are characterized by gross national products that are 10 to 40 times lower than average for developed countries. They have large inequalities in income distribution characterized by economic dependency, low levels of urbanization, deficient water sanitation, and poor transportation, communications, and technological development. Education levels are low, especially for women, and birth rates are high.[2]

Wide differences in economic and social development exist among resource-poor countries: Argentina, Mexico, and Brazil are quite different from Djibouti, Zimbabwe, or Bolivia. These differences are economic, cultural, and social. Diverse population groups exist within each country. For instance, wealthy populations whose illnesses and health problems are similar to those found in industrialized nations and who have access to private insurance and sophisticated medicine coexist with people who live in abject poverty without any access to health care. Because of these differences, generalizing about developing countries requires caution.

The World Health Organization (WHO) reports that the average life expectancy in underdeveloped countries is lower than in industrialized nations. In 2002, while life expectancy at birth reached seventy-eight years for women in developed countries, was less than forty-six years for men in sub-Saharan Africa.[3] Almost fifty-seven million people died world-wide in 2002, ten and one half million (nearly 20 percent) of whom were children younger than five years of age. Of these child deaths, 98 percent occurred in developing countries.[4] The main causes of death among children are depressingly recognizable: perinatal conditions closely associated with poverty, diarrheal diseases, pneumonia and other lower respiratory tract conditions, and malaria. HIV/AIDS, now the world's leading cause of death in adults between 15 and 59 years of age, is killing almost 5000 men and women in this age group, and almost 1000 of their children, every 24 hours in sub-Saharan Africa.[5]

WHO points out that low contraceptive usage and lack of access to safe therapeutic abortion are serious health issues in developing countries. Legal restrictions to induced abortion in some developing countries do not deter its practice, but instead make abortions unsafe, forcing women seeking abortions into an environment that has minimal medical standards and where persons lacking necessary skills perform the procedures. In developing countries, the risk of death due to complications from unsafe abortion procedures is several hundred times higher than that of abortion performed professionally under safe conditions. A woman living in sub-Saharan Africa has a 1 in 16 chance of dying in pregnancy or childbirth. This compares with a 1 in 2,800 risk for a woman from a developed region.[6] An estimated forty-six million pregnancies end in induced abortion each year. Estimates judge that nearly 20 million of these are done under unsafe conditions. About 13 per cent of pregnancy-related deaths, or 67,000 deaths annually, result from the complications of unsafe abortion.[7]

Forty-two percent of deaths at all ages in developing countries are potentially avoidable. These are caused by communicable diseases, maternal and perinatal conditions, and nutritional deficiencies. Of these, we see a marked contrast in patterns of health transition among adults, aged fifteen years and over, in different parts of the world.

In developed countries, communicable diseases and maternal, perinatal, and nutritional conditions contribute only 5 percent to the total burden of disease, while in high-mortality developing regions, this figure rises to 40 percent. In African regions where the HIV/AIDS epidemic has confounded the pattern of health transition during the past decade, these conditions can contribute as much as 50 to 60 percent of the overall disease burden.[8] The AIDS pandemic is currently ravaging some developing countries. An estimated 90 percent of new infections occur in developing countries, where the disease has already reduced life expectancy in some countries by more than a decade. As revealed in the Second World Conference on AIDS, in July 2003, in poor countries only 5 percent of the population receives adequate treatment.[9]

From these data we can infer that for populations living in the developing world, the main causes of morbidity and mortality are still linked to environmental factors correlated with poverty, infections, and malnutrition.

Many resource-poor countries are experiencing an epidemiological transition, with a relative increase in the role of genetic factors on health and disease.[10] Between 1960 and 1997, the average infant mortality rate in developing countries fell from 138 to 65 per 1000. This reduction was especially manifest in Latin American, the Caribbean, East Asia, the Pacific, the Middle East, and North Africa. This was not the case for Sub-Saharan Africa and South Asia.[11] The improving health indices in some developing countries are indicative of social and economic progress associated with improved health-care services and control of infectious diseases and malnutrition. Consequently, genetically determined disorders and birth defects have begun to constitute an increasing proportion of infant and general mortality and morbidity, and are gaining public health significance. In the 1990s, deaths from genetic disorders in Latin American countries ranged from 8 to 25 percent of the total number of deaths as compared to 23 percent in the United States. Similarly, the percentage of deaths due to cardiovascular diseases in this region ranged from 8 to 46 percent, which is approaching the 54 percent recorded in the United States.[12]

3. Biomedicine Today

The possibilities of biomedicine represent an unstable arena. For instance, among geneticists, strong disagreements exist about the value, impact, and short-medium term possibilities of genetics per se.

Positions range from those who see genetics in an optimistic light and locate its possibilities in the diagnosis of common diseases, and those that view genomic medicine as a new form of preventive medicine, which we will achieve in approximately twenty to thirty years, to the skeptics that minimize its future impact in the short or medium term. The optimistic and enthusiastic predict that we will move from testing only a few families for rare disorders, to a wider diagnostic paradigm that will widen carrier and susceptibility screening in only a few years. As the process to scan for mutations becomes faster and cheaper, genetic testing will become more widespread. The second view considers preventive medicine as an economic necessity and genomic medicine as the best route to achieve that, but thinks that we will need more than twenty-five years to change the way we do medicine.

Finally, the skeptic or cautious view holds that exaggerating the importance of genetic factors stops people from thinking about the need to clean up the environment and tackle socioeconomic inequity. Neil Holtzman has discussed this view. His quarrel is with those who exaggerate the clinical benefits that will accrue from the human genome project. For example, geneticists believe that application of knowledge from the project will eventually materially benefit almost everyone in the world. Holtzman describes claims such as these to be ludicrous. About pharmacogenomics, he says that sometimes testing patients for designated enzymes involved in drug metabolism may help, but to suggest that inherited genetic profiles wholly determine drug sensitivity and resistance is ridicuous.[13]

Similar disagreements arise about genetics in the context of resource-poor countries. To assess whether biomedicine is a real benefit for developing countries, let us first consider the current state of affairs in the field. Most biomedical therapies are developed in industrialized countries and then implemented in developing countries. Extrapolation from these data presents difficulties because of significantly different socioeconomic conditions prevailing in the countries where research was conducted and the developing nations where the results are applied.

A primary issue for consideration is the kind of diseases chosen as the target of research. Present research in molecular genetics has made great strides in understanding diseases that evince Mendelian inheritance, for example, as in the hemoglobinopathies. Currently, more than 600 disorders can be diagnosed by DNA analysis.[14] But even if we can speak of an epidemiological shift in developing countries, a prevalence of infectious diseases remains. Yet most research investigating genetics does not target infectious diseases.

In multifactorial illnesses, genes do not play a significant determinant role. These illnesses are due to the interaction of several genes, or only a genetic predisposition with environmental factors.

Even if we do not target the main causes of illnesses and mortality in developing countries, if we want these new technologies to be useful to them, we should shift from the individual perspective to a public health view. We should think about the ways that biomedicine, genetic tools, DNA recombinant technology, and assisted reproduction could improve the health or quality of life of populations that live in resource-poor countries.

Consider the case of chronic multifactor illnesses in adults, such as cancers and cardiovascular diseases. These represent a major burden for industrialized and developing countries. These diseases have a complex causality based on an interaction between genes and environment. We can design different strategies for investigation. For example, for cancer, one strategy can focus on environmental factors such as the control of tobacco usage, radiation, and contaminants, while another can identify individuals with a genetic predisposition. The Argentine geneticist, Victor Penchaszadeh, points out that even if genomics and protogenomics might help in some cases, on a population level, controlling environmental factors for genetically predisposed individuals would be more efficacious.[15]

Tobacco usage control is a good example due to its impact on health and the lack of policies to fight it in many developing countries. The latest epidemiological studies indicate that death rates for smokers are two or three times higher than for non-smokers at all ages. Smoking trends show that cigarettes consumption in developing countries has been rising among men over the last three or four decades. Smoking causes approximately half a million or more deaths per year in China. Per capita tobacco usage is falling in the developed countries at a rate of about 1.5 percent per year, but rising in developing countries at a rate of about 1.7 percent per year.[16] Some have suggested public campaigns and taxes on cigarettes as ways that might be effective to reduce tobacco usage.

Second, consider single-gene inheritance, which, though not frequent, does impact public health. This is the case for some hemoglobinopathies and other illnesses prevalent in some populations, such as Spinocerebellar Ataxia, common in Cuba,[17] or Huntington's Chorea common in Venezuela.[18] This is an area where genetics has had a significant impact, but the success has been in diagnosis instead of in prevention or treatment strategies. This creates a gap between diagnosis and therapy that poses a major ethical dilemma when no prevention or treatment exists. In countries that suffer a shortage of resources, having only the diagnosis is useless.

We could meliorate the problem of some serious inherited disorders though genetic screening for carrier identification. Based on screening results, we could offer genetic counseling, contraception, reproductive choices that include assisted reproduction that could screen out affected zygotes, prenatal diagnosis, and, in cases where afflicted fetuses are identified, abortion if desired. This apparently simple strategy can be quite complicated in some developing countries. The Argentine case is illustrative regarding the influence of social context in the implementation of strategies such as genetic prenatal testing, abortion, and assisted reproduction.

Consider these two facts:

(1) In 2003, problems still existed with the implementation of a National Program on Reproductive Health, designed to give out information and contraceptives in public hospitals.

(2) Argentinian law prohibits abortion.19 The Penal Code includes three exceptions, but conservative judges interpret them as only two: when the mother's life or health is in danger and when the pregnancy was caused by the rape of a woman with intellectual impairments. Judges will not accept abortion in the case of a raped woman. They read the law as if only mentally retarded rape victims may receive legal abortions. Even when the more conservative interpretation is adopted,

> and abortion is still approved, physicians tend not to perform abortions in public hospitals.

In Argentina, choice based on results of prenatal diagnostic testing is impossible. Abortion is illegal even subsequent to identification of fatal genetic conditions of the fetus such as anencephalic fetuses that will die at birth or some hours later. Armed with such a diagnosis, parents face the psychological and physical burden of knowing that they must endure the doomed pregnancy with no legal options open to them. The City of Buenos Aires passed a law in June 2003, which allows interruption of pregnancy after six months if the fetus is anencephalic. Notice that this only applies in one city of the Republic and can only be done after six months pregnancy even though ultrasound tests and genetic prenatal diagnoses are made quite early in pregnancy. Geneticists do not acknowledge this problem. They only focus on technique, ignoring the social and legal context.

Assisted reproduction poses similar dilemmas. On one hand, many assisted reproduction centers, while focused on assisting reproductive efforts, have ignored other problems related to the sociocultural context that causes secondary infertility such as lack of safe abortion, adequate contraception, and sex education; the high rate of teenage pregnancies; and a 46 percent hospitalization rate for abortion complications that produce infertility. Adequate sexual education, contraception, safe abortions, and treatment of sexually transmitted diseases can prevent secondary infertility. In these, social changes that respect the gender perspective should be seriously encouraged. On the other hand, despite the denial of assisted reproduction clinics, some critics also charge that these centers, in some cases, manipulate or discard viable embryos. The policy has been to introduce sophisticated assisted reproduction techniques without a deeper commitment to reproductive health or satisfactorily addressing ethical issues.

In many countries, just extrapolating the technology that works in industrialized countries is not enough. There should be a broader view to have a rational and useful introduction of technology that meets the needs of the context into which the new technology is introduced.

This is not to say that we should not transfer techniques, or that techniques developed in industrialized nations should be banned in developing countries. My position is to encourage caution about how to do this and to propose that we consider the potential impact of these technologies in different sociocultural contexts. Needs in developing countries may not be the same, and the causes and solutions of problems may be different in these countries from those in industrialized nations.

4. Biomedicine Research in Developing Countries

Given the morbidity and mortality in resource-poor countries, we need to think of genetics or biomedicine in new, nontraditional ways. Until now, the primary focus of genetics has been on rare diseases and non-communicable diseases, mostly those that affect developed nations: cancer, and pathologies such as Alzheimer's and Parkinson disease. Developing countries suffer most from infectious diseases, exacerbated by the lack of infrastructure (lack of electricity, impure water, and transportation problems) and social conditions such as the oppression of women. Lack of potable water, for example, fosters diarrheas and other illnesses; lack of electricity prevents adequate refrigeration of foods and medicines; transportation problems hampers access to medical centers and physicians; and social oppression of women poses obstacles to obtaining adequate care for mothers and their children.

If biomedicine is to be effective, and we are to think of it as worthwhile for developing countries, then we should redirect its priorities to tackle the causes of these diseases prevalent in developing countries. We should think of biomedicine from a public health perspective and from the social context and infrastructure where it will be applied. We need research on parasites and illnesses prevalent in developing countries, and medicines or cures that do not depend on refrigeration or on several visits to the clinics. We need to explore not only innovative ways of prevention and cure, but also ways to provide access to those interventions.

Let me present a study done on Chagas disease as a case example. Chagas is a chronic, frequently fatal, disease endemic in Central and South America. An estimated sixteen to eighteen million persons are infected with *Trypanosoma cruzi*, the parasite that causes Chagas, transmitted by blood-feeding bugs such as *Triatoma infestans*, *Rhodnius prolixus*, and *Panstrongylus megistus*. One hundred million people, roughly one quarter of the population of South America, are at risk of infection. Two drugs are curative in the acute and early chronic phase of infection but have adverse side effects and may not eliminate *Trypanosoma cruzi*. But neither adequate drugs nor vaccines are available. Some researchers are studying the molecular biology and genome of the parasite. Unfortunately, to date, results have not been promising.

Meanwhile, a sophisticated ecological study published in *Science* examined alternative control strategies. This study used a mathematical model calibrated to detailed household data from three rural villages in northern Argentina.[20] The study showed that householders could greatly reduce the risk of human infection by excluding domestic animals, especially infected dogs, from bedrooms. The results are eloquent: low cost, locally practicable, environmental management. The science needed to perform the study was quite sophisticated, but the answer provided is simple. Educational programs aimed at changing behavior of householders can significantly meliorate the disease.

If we are to redirect science to focus on concrete problems of developing countries, we need to address some salient questions. Does such a shift present a relatively promising manner in which to conduct research on infectious diseases and on other major health problems for resource-poor countries? Does it present new technological ways to prevent these problems? If we agree that such a shift is a promising scientific possibility, is it technically feasible? Is it realistic to expect that the shift occur given the present global socioeconomic situation of health and research?

Abdallah Salim Daar and Peter A. Singer express optimism with regard to these questions. In "Top Ten Biotechnologies for Improving Health in Developing Countries," they identify priority technologies.[21] They conducted a foresight study with a panel of twenty-eight experts. The criteria used by the panelists for their assessments were impact: how much difference a technology will make in improving health; appropriateness: how affordable, robust, and adjustable to health-care settings the technology could be; burden: whether it addresses the most pressing needs; feasibility: whether the technology realistically be developed and deployed in a time frame of five to ten years; knowledge gap: whether the research creates new knowledge; and indirect benefits: whether the technology addresses environment improvement and income that have indirect, positive effects on health. Based on these criteria, experts listed the top three technologies as modified molecular technologies for affordable, simple diagnosis of infectious diseases, recombinant technologies to develop vaccines against infectious diseases, and technologies for a more efficient drug and vaccine delivery system. Examples of the first are polymerase chain reaction (PCR), monoclonal antibodies, recombinant antigens, and PCR-based HIV tests that detect the pro-viral DNA in infants with the use of filter paper to process and store blood samples (these are heat stable and usable for many months). The authors point out:

> Simple hand test-devices that rely on the binding specificity of monoclonal antibodies or recombinant antigens to diagnose infections may be easily *adaptable to settings without running water, refrigeration, or electricity.*[22]

Examples of the second are a malaria vaccine and a recombinant hepatitis B vaccine. In the third category, the authors noted that most vaccines and many drugs are administered by injection, and that tens of thousands of new cases of blood-borne diseases, such as HIV/AIDS and hepatitis B, are caused by unsanitary injections. They pointed to the enormous expense of refrigeration—maintaining required temperatures can add up to 80 percent of the cost of vaccine delivery in resource-poor countries—and the inconvenience of frequent dosing as two drawbacks to current methods of vaccine and drug delivery. The alternatives suggested include powdered, edible vaccines and controlled-release formulations that replace the need for multiple doses.[23]

Daar and Singer also mentioned other relevant techniques such as those for environmental improvement (sanitation, clean water, bio-remediation), female-controlled protection against sexually transmitted diseases with and without contraceptive features, genetically modified crops with increased nutrients to counter deficiencies, and recombinant techniques to make therapeutic products such as insulin and interferon more affordable. In addition to the heterogeneity of these techniques, each has different possibilities and all are in different stages of development.

While, work in recombinant techniques for interferons or eritropoietin has been ongoing for some time, developing new vaccines is much more difficult because this work needs large investments of funding over a protracted period to cover all aspects of the work, including patent costs.

For other proposals such as bio-remediation, technical issues are not the problem. In many cases, we already have developed technology that cannot be implemented because of political or economic reasons. For example, we have long known how to produce clean potable water. For these countries, instead of technical challenges, the problems are economic or political. Regardless of technical expertise, these countries lack adequate funding and the political force to direct the funding. Unless the affordability projection is promising, with no risk to the environment, investing existing scarce resources in this kind of research is not a high priority for the developing world.

Daar and Singer also analyzed frequent assumptions about the applicability of biotechnology in developing countries. They challenge the concept that biotechnology is irrelevant to the world's poor, and show that biotechnology can contribute to the prevention of disease and the promotion of health.

If such research is sound and has serious possibilities, then my first question has an affirmative answer: We can find relatively promising ways to conduct research on treatments for, or ways to prevent, infectious diseases and illnesses prevalent in developing countries.

Let us consider whether we will see this research conducted, given the present global socioeconomic situation of health and research. Several issues need attention.

The first concern addresses a need to redirect the focus of research. Who will be interested in financing this kind of research? The pharmaceutical industry will have little interest because these are countries with few financial resources to buy their products. The resource-poor countries will have to bear the financial burden. Differences among developing countries are critical. In some countries, such as Bolivia and Paraguay, this shift in economic resources would be nearly impossible, while other countries, such as Argentina, Brazil, and Cuba, that have some scientific means, could adopt a sound policy on research. Even in developing countries with the capacity to do research, frequently social, economic, or cultural conditions that impede local efforts. These conditions include lack of research funding, lack of continued and coherent policies on national research goals, and government corruption that prevents development of research infrastructure.

Besides these local problems, academic incentives may conspire against a redirection in science. Normally academia and research institutes reward publications in prestigious international journals, or new patents. Many of the efforts we want to promote are not innovative enough to be published or patented. They are the product of the application of sophisticated science, but are not scientifically original. If a researcher wants to publish in top international journals, normally he or she will work in similar topics to those studied in industrialized nations. To encourage academics to redirect their focus, the developing country will have to set different parameters and criteria than the ones in the central countries. This is another barrier. Even if resource-poor countries should support and promote scientific efforts applied to their problems, given the resistance of the academia, the urgent needs, and the lack of scientific policies, the task for them is most difficult.

Even if a redirection of research may be desirable, it may be quite complicated to achieve. International agencies such as the World Health Organization or other international initiatives may be possible sources of funding, but few organizations such as these exist, and these will not be able to support all the needed research. A possible solution may be the promotion of this new agenda by international agencies and the encouragement of regional and south-south collaboration. For instance, regarding the *Tripanozoma cruzi*, Argentine research teams are currently working with South African teams.

The second concern has to do with the accessibility of diagnostic or treatment techniques. We see a long history of research findings that have not been accessible to the populations that needs them most. Reider Lie provides the example of the hepatitis A vaccine research done in Northern Thailand in 1991.[24] That study involved approximately 40,000 children between one and sixteen years old. Lie stated, "The hepatitis A vaccine has been licensed by Smith Kline Beecham, and is now mainly used for people who travel from industrialized countries to developing countries." He continued:

> The main reason for doing the trial was to have the vaccine licensed in industrialized countries. The Thai population was a convenient one, with a high incidence of hepatitis A. This trial then is an instance of a trial *done in a developing country, in the knowledge that the product would not be available to the population after testing, and that the results will primarily benefit people in the developed world.*[25]

An expensive vaccine for only one year of immunization is not cost effective for developing countries.

In light of past events, we must inquire whether these techniques will be affordable and available. Singer and Daar suggest that they will. They believe that we can make biotechnology, especially molecular diagnosis, affordable for developing countries. Researchers consulted agree with regards to diagnostic kits.[26] The new diagnostic methods need a quite sophisticated and complex technology, but the necessary equipment is not difficult to build, so normally patent costs for equipment are not an issue, but new drugs, proteins, or vaccines probably will be patented. Vaccines need large investments in research for long periods. Singer and Daar point out that "enforcement of intellectual property rights will be crucial to the affordability of these technologies" and where the interests of developing countries are concerned, "key stakeholders can be magnanimous." Even if magnanimity may be a starting point, to modify the health situation of these countries, random charity is insufficient. We should move into the realm of rights and obligations.[27] Gaps among populations in the world are widening and the help and charity from industrialized countries are still scarce.

If these scientific efforts are to be undertaken, they should be done in a way that allows availability to populations who need them. Again, a possibility may be the help or coordination of non-profit international agencies such as WHO. These organizations could hold the patents or negotiate prior agreements to make drugs or vaccines available. WHO has already made prior agreements regarding malaria, schistomiasis, and contraceptives. Though promising, and maybe one of the few workable options, this still has limits. Notice that in December 2003, the recently appointed Director General Lee Wook Jong promised a program to combat AIDS. Its aim is to provide treatment for three million patients in the year 2005—a low target when the estimate of persons infected in the world is forty-two million!

5. Conclusion

Evaluating these new biomedical technologies for resource-poor countries is quite difficult. In this chapter, I considered the present situation of illnesses in developing countries and the impact of existing biomedicine in these countries. Then I analyzed possible techno-scientific efforts.

The main factors that explain the persistence and worsening of infections such as tuberculosis, malaria, or AIDS are social, economic, and cultural. Unless we seriously evaluate and address these factors, no amount of technology will fix them. Known methods such as improving the infrastructure to provide pure potable water, a hygienic sewage system, good nutrition, and access to health care can prevent a significant percentage of health problems in resource-poor countries. These elements are basic. We must not lose sight of priorities.

The second issue concerns the adoption of already existent biomedicine. We must consider the health problems of each country adopt relevant tools instead of merely extrapolating what has been developed for industrialized countries. We have to avoid creating false needs. Not all technological devices are needed or relevant to needs. The economic and social context in the target countries is relevant and we should seriously consider it to identify and implement effective treatments.

Third, science and research may be useful for some technological or scientific problems, such as quick and affordable diagnostic tests for early treatment or vaccines for infectious diseases. The way we currently do science, we prioritize the needs of industrialized countries over those of developing countries. We should redirect science to focus on the problems prevalent in developing countries. Even if we acknowledge potential benefits of this new research, resource-poor countries will have considerable difficulty achieving the benefits independently. A possible solution may emerge from international initiatives or agencies.

Undoubtedly, while we can use science to solve problems and help design adequate tools, in many cases, scientific research is a necessary, but not sufficient, condition for development. For many decades, we have known how to solve problems such as the need to have running water, electricity, and sanitation. But if the poor do not have access to that knowledge, or do not have the financial resources to apply that knowledge to practical solutions, then deprivation will continue.

NOTES

1. United Nations Development Program (UNDP), *Human Development Report 1998* (New York and Oxford: Oxford University Press, 1998).

2. World Health Organization, World Alliance of Organizations for the Prevention of Birth Defects, *Services for the Prevention and Management of Genetic Disorders and Birth Defects in Developing Countries: Report of a Joint WHO/WAOPBD Meeting, the Hague, 5–7 January 1999* (Geneva: World Health Organization, Human Genetics Programme, 1999).

3. World Health Organization. *The World Health Report 2003: Shaping the Future* (Geneva: World Health Organization, 2003), chap. 1.

4. Ibid.

5. Ibid.

6. Carla Abou Zahr and Tessa M. Wardlaw, *Maternal Mortality in 2000: Estimates Developed by WHO, UNICEF, and UNFPA* (Geneva: Department of Reproductive Health and Research, World Health Organization, 2004).

7. World Health Organization, *Safe Abortion: Technical and Policy Guidance for Health Systems* (Geneva: World Health Organization, 2003).

8. World Health Organization. *The World Health Report 2003*, chap. 1.

9. Czubaj, F. "*Segunda Conferencia Mundial sobre Sida: Sólo el 5 Percent de los Infectados Recibe Atención Adecuada, Unos 5000 Especializtas Debaten en París*," ("Second World-Wide Conference on AIDS: Only 5 Percent of Those Infected Receive Suitable Attention, About 5000 Specializts Debate in Paris"), *La Nación* (15 de Julio 2003), p.10.

10. World Health Organization, World Alliance of Organizations for the Prevention of Birth Defects, *Services for the Prevention and Management of Genetic Disorders and Birth Defects in Developing Countries*, p. 3.

11. Zahr and Wardlaw, *Maternal Mortality in 2000.*

12. V. Penchaszadeh, "Genética Humana y Salud Pública," *Ciencia e Saude Colectiva*, 7 (2002), pp. 33–36, esp. 34.

13. Tessa Richards, "Three Views of Genetics: the Enthusiast, the Visionary, and the Sceptic," *British Medical Journal*, 322 (28 April 2001), p. 1016.

14. Penchaszadeh, "Genética Humana y Salud Pública," p. 3.

15. Ibid., p. 4.

16. S. Jabbour, Reddy K. Srinath, W. F. T. Muna, and A. Achutti, "Cardiovascular Disease and the Global Tobacco Epidemic: A Wake-Up Call for Cardiologists," *International Journal of Cardiology*, 86 (2002), pp. 185–192.

17. G. Auburger, Diaz G. Orozco, Capote R. Ferreira, Sánchez S. Gisper, et al., "Autosomal Dominant Ataxia: Genetic Evidence for Locus Heterogeneity from a Cuban Founder-Effect Population," *American Journal of Human Genetics*, 46 (1990), pp. 1163–1177.

18. R. Avila-Giron R, "Huntington Chorea 1872–1972," A. Barbeau, T. N. Chase, and G. W. Paulson, eds., *Advances in Neurology*, 1 (1973), pp. 261–266.

19. Penal Code of Argentina, Article 85.

20. J. E. Cohen and R. E. Gürler, "Modeling Household Transmission of American Trypanosomiasis," *Science*, 293 (2001), pp. 694–698.

21. A. Daar and P. A. Singer, "Top Ten Biotechnologies for Improving Health in Developing Countries," *Nature Genetics*, 32:2 (October 2002), pp. 229–232.

22. Ibid., p. 230 (emphasis added).

23. Ibid.

24. B. L. Innis, et al., "Protection against Hepatitis A by an Inactivated Vaccine," *Journal of the American Medical Association*, 271 (1994), pp. 1328–1334.

25. R. Lie, "Justice and International Research," *Biomedical Research Ethics: Updating International Guidelines: A Consultation: Geneva, Switzerland, 15–17 March 2000*, eds. Robert J. Levine, Samuel Gorovitz, and James Gallaher (Geneva: CIOMS, 2000), pp. 30 (emphasis added).

26. Lino Barañao, President of the National Agency of Scientific and Technological Promotion of Argentina (*Agencia de Promoción Científica y Tecnológica*), Alberto Kornblit, Researcher at the National Research Council of Argentina (CONICET), and Patricia Saidón, Coordinator of Drugs Evaluation Area of the Argentine Agency of Drugs and Medical Devices (ANMAT), personal communications (September 2002).

27. Florencia Luna, "Poverty and Inequality: Challenges for the IAB," Presidential Address delivered to the Seventh World Congress of Bioethics (November 2004), Sydney, Australia, published in *Bioethics*, 19:5–6 (October 2005), pp. 451–459; and Thomas Pogge, *World Poverty and Human Rights* (Oxford: Polity, 2002).

Four

BRAZILIAN RESEARCH ETHICS: A NORTH-SOUTH DIALOGUE AIMING TO BUILD A NEW CULTURE OF RESPECT

Leo Pessini and Leonard M. Martin

1. Introduction

One of the characteristics of Brazil as an emerging nation is its growing competence and confidence in the area of scientific and technological research. Considerable advances have occurred in the fields of pharmaceutics, medicine, and the life sciences in general. As a result, we have seen a growing awareness of the need for ethical oversight of research done on human subjects to prevent exploitation of poor individuals, victims of cultural and economic inequalities, who survive on incomes below subsistence level. These people are vulnerable due to cultural and economic inequalities

The aim of this article is to trace briefly the emergence of Brazilian research ethics in recent years, showing that this phenomenon is at once the result of challenges presented by the local reality and of dialogue with first world bioethics. We will show that, while learning from the bioethics of developed countries and from different international declarations and documents, the institutionalization of research ethics and the proliferation of committees for ethics in research have created a forum in which ethical discussion with a uniquely Brazilian flavor has emerged and is flourishing. This interaction between local needs and challenges and theoretical reflection originating in the United States and Europe has produced what we would hope to characterize as a fruitful North-South Dialogue.

2. The Emergence of Brazilian Research Ethics

Because of Brazil's colonial history, only after gaining independence in the early nineteenth century did third level institutions begin to emerge in the country. Universities only began to appear and flourish in the twentieth century. Consequently, research of all sorts, including biomedical research, got off to a slow start. By the second half of the twentieth century, Brazilian research capacity was beginning to thrive and with it, a growing interest in ethical issues.

We find early signs of concern with the ethics of research with human subjects in codes of medical ethics that Brazil has produced with a regularity.[1]

The 1953 Code is one of the first to devote space to the issue, adopting an attitude of hesitancy with regard to experimenting on human beings. Article 57/1953 delimits the practice, condemning experiments in what it terms *anima nobili* (on human beings) for purely speculative purposes, even if investigators obtain consent. The code only tolerates those experiments which have a strictly therapeutic or diagnostic purpose, carried out in the interest of the sick person, and which, with all certainty, are not life threatening or capable of causing serious damage to the subject.

Reluctance to endorse research on human beings is partially the result of the still-fresh memory of atrocities committed during World War II. This sentiment also explains the blanket prohibition in Article 58/1953 against any experiments whatsoever for purposes of war, politics, race, or eugenics, a prohibition that reappears in later codes (Article 32/1984 and Article 122/1988).

From the beginning of the Brazilian tradition of codified ethics, as far as research is concerned, human rights and dignity are valued more highly and take precedence over any scientific interest or concern.

The code published in 1984 takes a less cautious approach to the general practice of human experimentation, recognizing that research in human beings for scientific purposes is legitimate, but it does insist that such research must have proper supervision. Article 30/1984, for example, has a requirement that experiments be properly authorized and by an ethics committee. Although another ten years or more passed before these requirements to become effective in Brazil, their appearance in official documents was significant.

The 1984 Code addresses social control over the issue of consent.[2] Article 31/1984 forbids a doctor from using any type of experimental treatment that has not yet been approved for use in the country, without first obtaining authorization from the competent authorities. The article also requires consent be obtained from patients or their legally guardians after they are provided adequate information about the condition of the patient and possible consequences of the experimental treatment. We find the same prohibition in Article 124/1988.

The Brazilian *Code of Medical Ethics*, published by the Federal Medical Council in 1988, is currently in force. The growing interest in research ethics is reflected in the space dedicated to the subject of ethics: an entire chapter comprising nine articles on the topic.[3] The conviction that human beings have intrinsic value and bear universal and inalienable rights, runs though the document. This conviction repudiates depersonalization and consequent instrumentalization in the service of any ideology. The code reflects the value placed on the human person in its treatment of consent and its appeal to the principles of autonomy and beneficence. Article 123/1988 forbids experiments on human subjects unless researchers obtain written, informed consent given after subjects receive adequate information about the nature and consequences of the research project. An additional paragraph within the same article states that if the potential subject has diminished or no capacity, experimentation can only be conducted with the express authorization of the person legally responsible for the patient and, even then, only if the incapacitated person can be expected to benefit from participation. The code abhors the idea of using such a person merely as an instrument to benefit science.

The emphasis on written consent following a process of information and explanation has the goal of guaranteeing liberty and avoiding manipulation or depersonalization of research subjects. Genuine communication between the researcher and the subjects is encouraged. Here we have early signs of thinking that ideally there should exist between the researcher and the person being researched a relationship of partnership, however unequal that relationship might be.

Extending the code of 1984, the 1988 code mandated increased social control by requiring that research be properly supervised. Article 127/1988, elaborating on Article 30/1984, affirms that before medical research may be conducted using human beings, investigators must first submit a research protocol to a commission completely independent of the researcher. This requirement, more than merely a bureaucratic measure, has a double function. It guarantees the scientific soundness of the project and ensures its ethical consistency. A scientifically unsound research project is, by nature, unethical, because it poses risks to the human research subjects. Ethics committees can appreciate and better evaluate studies that explicitly elaborate the research protocol.

The 1988 code evinces awareness that social control is a key factor when proposed research has a social or communitarian dimension. Article 125/1988 mandates that a people be informed that research is being done in their community. In addition, researchers must adopt the objective to protect public health, and they must respect local customs and conditions. While this article has general application, it has special relevance to research projects involving Brazilian indigenous populations, often extremely vulnerable because of cultural disparities. Awareness of these complexities are reflected in National Health Council resolutions which provide explicit protections and safeguards.[4]

The growth of the techno-scientific paradigm of medicine has been a consequence of scientific and technological developments in the country. In response, Brazilian research ethics has emphasized that science and medicine have made enormous contributions to the world, they should exist to serve humanity; human persons do not exist to serve science and medicine. The codes have appealed to the principles of autonomy and beneficence to protect the dignity of human beings used as research subjects.

Research ethics has also addressed issues concerning research funding. Scientific and technological medicine is expensive and requires huge financial investments, which come from state sources and the private sector. The growing weight of the economic factor has led to the emergence of the commercial-entrepreneurial paradigm of medicine where the profit motive has a strong, if not preponderant role. In the area of research, this paradigm has an especially powerful influence because of the huge budgets invested in, for example, pharmaceutical research, in the hope of reaping enormous profits through the sale of new, more effective drugs. Brazilian research ethics does not oppose economically sound medicine and research any more than it opposes scientifically sound medicine, but it questions subordinating the interests of the human being, who is the subject of research, to the interests of profit and of big business.[5]

Alongside concerns such as social control and consent, a strong feature of Brazilian research ethics is its preoccupation with vulnerable individuals. Article 128/1988 for example, forbids medical research in volunteers, healthy or not, who are in any way dependent on, or subordinate to, the researcher. The intention is to protect, for example, students and employees in situations where their liberty to refuse participation may be limited, from being exploited either by their professor or their employer. We can identify parallel categories of vulnerability in prison populations and military personnel.

We see concern with vulnerability reflected in other parts of the code. Article 129/1988 prohibits implementing procedures or conducting medical research that suspends or omits standard therapeutic practice if, in doing so, a patient would be harmed. The code stipulates that the good of the patient must take precedence over the interests of science. Article 130/1988 prohibits experiments using new clinical or surgical techniques on incurable or terminally ill patients unless reasonable hope exists that the procedure might be of benefit to the patient and patients endure no additional suffering.

Also in 1988, the Brazilian National Health Council included ethical concerns in the elaboration of research protocols that involve experimentation on human subjects with the publication of Resolution 01/88.[6] Unfortunately, this document did not have much impact—despite its emphasis on respect for human dignity—on the importance of informed consent and on the need for independent ethics committees.[7] In 1995, the National Health Council created a working party to review the 1988 resolution. As a result of their efforts a new text emerged, Resolution 196/96, which was promulgated in October of 1996.[8] The 1996 resolution has become the standard reference for human research ethics in Brazil, enhanced by a series of other resolutions designed to increase its effectiveness.

The main difference between the 1996 resolution and its predecessor is the creation of structures to implement its norms explained the rationale for the norms. As a direct result, Brazil established the National Commission for Ethics in Research (CONEP) and, at the same time, a network of Research Ethics Committees (RECs) distributed throughout the country.[9]

3. Learning from Developed Countries and International Declarations and Documents

Brazilian research ethics has never been shy about learning from foreign sources, whether international declarations and documents, or local sources of different kinds, including the writings of bioethicists.[10] One of the features of dialogue is the ability to listen others and learn from experience. Brazilian texts such as the *Code of Medical Ethics* and Resolution 196/96 acknowledge their debt to documents such as the Nuremberg Code of 1947[11] and the Declaration of Helsinki, originally put forth in 1964 and revised several times since then, most recently in 2000.[12]

Resolution 196/96 explicitly recognizes as part of its foundation, the principalist ethics characteristic of North American bioethics in its formative years.[13] Elements assimilated from international sources include emphasis on informed consent and the liberty of the research subject to elect or refuse participation; insistence that experiments on human subjects be done according to sound scientific principles formulated in research protocols that are submitted to a properly constituted REC for approval before any intervention occurs; and the principle of nonmaleficence, the duty to do no harm. In therapeutic research, where some measure of harm to the patient is unavoidable, the postulated harm should be proportional to expected benefits. Risks and benefits need to be calculated, and the danger of death or permanent maiming of research subjects is unacceptable. To do no harm is not enough. Research involving human beings should positively have a reasonable perspective of producing beneficial results, if not for the subject directly—as in some clinical trials using healthy subjects—for society. Other features assimilated from external sources are concern for vulnerable populations and a commitment to justice.

4. The Brazilian Contribution to the Discussion

Dialogue involves learning from another and offering something in return. We will conclude this article looking Brazilian contributions to research ethics. A good starting point is CONEP, which serves as an example of how a country can overcome its limitations and create a climate in which it takes ethics seriously through generating mechanisms that guarantee decentralization and widespread participation in ethical reflection on research projects.

Resolution 196/96, which created the Commission, also has a variety of features that contribute to the North-South dialogue with significant contributions to global bioethical developments. The first such feature is that, although the text has a solid juridical basis reflecting the growing interface between law and bioethics, its primary nature is ethical. It regards the dignity of human persons as a central concern, discusses the fundamental principles that ought to guide research with human subjects, and invites people to internalize these values. The resolution addresses principles such as autonomy, beneficence, nonmaleficence, justice, and equity at length. The resolution devotes ample space to questions about the risks and benefits involved in research and free, informed consent. It strives to establish an independent formulation for these issues within the context of knowledge that Brazil has learned from the experience of others.

That the text deliberately sets out operative definitions of ethical norms that it proposes, prescribing mechanisms and structures for concrete action, will prove useful. Two such mechanisms, which receive detailed treatment, are the research protocol and the REC. Insisting on a properly constituted protocol forces researchers to articulate, in writing, exactly what they propose to do and to reflect carefully on the technical, scientific, and ethical aspects of the procedures they will use. A well-written protocol permits RECs to execute technical and ethical analysis of the research project.

One of the reasons why the 1988 resolution had so little impact was the lack of operational mechanisms. The 1996 resolution avoided this error by detailing the composition of the committee and its mandate, attributes, and organizational structure, and basic procedures. While respecting local autonomy and the principle of decentralization, the committees are not isolated one from the other. Under the coordination of CONEP, the resolution established a network of ethics committees, which provide mutual support and exchange of information, by means of regional and national meetings. An Internet website and a print periodical, *Cadernos de Ética em Pesquisa*, published several times per year by the National Commission supports intercommittee communication.

That currently almost 400 committees for research ethics are scattered throughout the whole of Brazil attests to the success of this marriage between theoretical concerns and practical measures. The National Commission's monthly meetings in Brasília and their employment of a permanent, full-time, executive secretariat are significant achievements. The secretariat does a great deal of the bureaucratic work involved in examining research protocols at the local and national levels, and is a significant link in the communications chain.

An element of Resolution 196/96 applicable beyond the borders of Brazil is the broad definition of research with human subjects adopted by the 1996 resolution, that being: "any research which, individually or collectively, involves human beings, directly or indirectly, in its totality or in its parts, including the manipulation of information or material."[14] The document aims at regulating all research in human beings and not just biomedical research. Within this perspective, sociological and anthropological research should also pass through the examination of independent ethics committees that would examine their ethicality.

Resolution 196/96 also admirably defends research subjects' interests. The fundamental assumption is that the interests of research subjects have higher priority than the interests of researchers do. Reciprocally, researchers' interests are best served by promoting respect for human subjects' dignity. From this perspective, informed consent is more than a legal document limiting researchers' responsibility or holding sponsors harmless for damages that may arise from the research project. Instead, informed consent in this context is viewed as an instrument that formalizes the process in which research subjects are informed, in unambiguous, easily understood language, of the objectives of the research project, the methods that will be used, the risks and discomforts that may be involved, and the benefits which may accrue. Based on this information, potential subjects can make free, informed decisions whether to participate. This personalized partnership approach to the consent process goes beyond the purely legal model, seeking, in respectful dialogue, to guarantee the dignity of all parties involved.

A third significant element of Resolution 196/96 is its belief in the benefits of social control over the area of research conducted with human subjects. Whether in developed or in developing countries, a great deal of the resistance to elaborating protocols and having to submit them to independent committees for ethical evaluation is based on the idea of self-regulation. According to this view, researchers believe that they do not need supervision by extrinsic to the research process. They especially object to oversight by non-scientists who are technically unprepared to appreciate scientific projects.

The long history of abuse of human rights that has plagued biomedical research is one of the strongest arguments against self-regulation. Scientific competence is not now, nor has it ever been, a guarantee of ethical competence or commitment to ethicality. Regardless of the technical competence of researchers, we believe that ethical issues require independent oversight to protect human subjects from harm or exploitation.

One of the unique contributions of Brazil to this debate is the proposal to guarantee social control by establishing multidisciplinary RECs where no single profession has a monopoly on membership. These committees would include men and women, and have representatives of a variety of professions including philosophers, theologians, bioethicists, lawyers, and representatives of the consumers of medical services that all serve with equal status. Although these multidisciplinary committees exercise effective social control, the mentality that permeates this ethical regulating of research is not one of repressive policing. On the contrary, the emphasis is on openness and promoting dialogue, giving preference to education, and trying to convince by argument instead of adopting repressive measures and sanctions, though such are available as a last resort.

5. Conclusion

In some quarters, the idea that developing countries are backward and can have little of interest to offer to any discussion is tempting. Especially in the area of ethics, we question this supposition. Exchange and cooperation can be fostered if develop partnerships. Partnership does not imply equality between parties, but it does imply dialogue and mutual respect, and the recognition that weaker parties may often be in the position to make valuable contributions to the understanding and resolution of ethical challenges. Economic power and scientific competence are not synonymous with ethical sensitivity. One of the challenges facing us is to develop a global mentality whereby human beings are not subordinated to the demands of scientific knowledge and economic power, but where science and economics are put at the service of human beings.

Open dialogue among professionals across North, Central, and South America is of fundamental importance. Those in South and Central America can contribute to bioethical reflection by developing a distinctive style of research ethics. This does not necessarily mean total originality. Developing countries are eager to learn from, and sometimes copy precedents set in the United States or in Europe. But increasing self-confidence and a growing awareness of self-worth in some developing countries have given rise to a new consciousness that participation in scientific and ethical advances is not limited to reproducing what has already been accomplished in the developed world.

Even when significant inequalities exist because of economic and cultural factors, all stakeholders can participate in meaningful dialogue respectful of their dignity and autonomy.

Developed and affluent societies must be vigilant to the implications that cultural differences evince regarding fundamental human values held by stakeholders in developing countries. Being economically vulnerable does not make a people bereft of humanizing values and practices. They can often offer perspectives, which can be refreshingly humanitarian in a culture saturated by the concerns of commercial and scientific medicine. In this spirit, we offered by way of example, the Brazilian experience in research ethics, where dialogue between the bioethics of developed countries and the concerns of the local culture has produced a dynamic body of practice and theory. The protocols developed by Brazil may be illustrative of ways that people in other developing countries might focus their research efforts. The Brazilian experience might also be enlightening even to those who already have quite sophisticated mechanisms of social control established.

NOTES

1. Cf. Leonard M. Martin, *A Ética Médica Diante do Paciente Terminal: Leitura Ético-teológica da* Relação *Médico-Paciente Terminal nos Códigos Brasileiros de Ética Médica* (*Medical Ethics Confronting the Terminal Patient: An Ethical-Theological Reading of the Physican-Patient Relationship in the Brazilian Codes of Medcial Ethics*) (Aparecida, Brazil: Editora Santuário, 1993), with special reference to the appendix which includes the text of all the Brazilian codes of medical ethics.

2. Len Doyal and Jeffrey S. Tobias, eds., *Informed Consent in Medical Research* (London: BMJ Books, 2001); and Joaquim Clotet, José Roberto Goldim, and Carlos Fernando Francisconi, eds., *Consentimento Informado e a sua Prática na Assistência e Pesquisa no Brasil* (*The Importance of the Informed Consent in Brazil in the Health Care and in the Field of Research*) (Porto Alegre, Portugal: EDIPUCRS, 2000).

3. Federal Medical Council, "Código de Ética Médica," (Code of Medical Ethics, 1988), Martin, *A Ética Médica Diante do Paciente Terminal*, pp. 389–399.

4. Conselho Nacional de Saúde "Resolução 304/2000 de 10 de Agosto de 2000" (Resolution 304/2000 of 20 August 2000), *Cadernos de Ética em Pesquisa*, 3:6 (November 2000), pp.25–27.

5. Cf. Leonard M. Martin, *The Human Rights in the Codes of Ethics of the Brazilian Medical Association* (São Paulo, Brazil: Loyola, 2002).

6. Conselho Nacional de Saúde (National Health Council) "Resolução 01/88") Resolution 01/88) in *Bioética—Revista do Conselho Federal de Medicina* (*Journal of the Federal Council of Medicine*), 3:2 (1995), pp. 137–154.

7. Cf. Carlos Fernando Magalhães Francisconi, Délio José Kipper, Gabriel Oselka, Joaquim Clotet, and José Roberto Goldim, "Comitês de Ética em Pesquisa—Levantamento de 26 Hospitais Brasileiros," ("Ethics Committes in Research: A survey in 26 Brazilian Hospitals") *Bioética—Revista do CFM*, 3:1 (1995), pp. 61–67.

8. Conselho Nacional de Saúde "Resolução N 196/196 de 10 de Outubro de 1996" ("Resolution No. 196/196, 10 October 1996"), *Cadernos de Ética em Pesquisa*, 1:1 (July1998), pp.34–42.

9. "Registros de CEP (Comitês de Ética em Pesquisa) (Registered Ethics Committees in Research), Instituições que Solicitaram Registro de CEP na CONEP até 18/06/99, por Estados da Federação" ("Institutions that Asked for Registration of the Ethics Committee in Research in CONEP until 18/06/99, by State of the Brazilian Federation"), *Cadernos de Ética em Pesquisa*, 2:3 (July 1999), pp. 25–30; Cf. "CEPs aprovados em 2001" ("CEPs approved in 2001"), in *Cadernos de Ética em Pesquisa*, 4:8 (Agosto 2001), p.30.

10. Cf. Leonard M. Martin, "*Ética em Pesquisa: uma Perspectiva Brasileira*" ("Ethics in Research: A Brazilian Perspective"), *O Mundo da Saúde*, 26:26 (January–March 2002), pp. 85–100.

11. Cf. "Nuremberg Code,," *Encyclopedia of Bioethics,* rev. ed., vol. 5, ed. Warren T. Reich (New York: Macmillan, 1995), pp. 2763–2764; and "The Nuremberg Code," *Informed Consent in Medical Research*, eds. Len Doyal and Jeffrey S. Tobias (London: BMJ Books, 2001), pp. 3–4.

12. Cf. World Medical Association, "Declaration of Helsinki: Recommendations Guiding Physicians in Biomedical Research Involving Human Subjects" *Informed Consent in Medical Research*, rev. ed., eds. Len Doyal and Jeffrey S. Tobias (London: BMJ Books, 2001), pp. 4–6; cf. World Medical Association, "Declaration of Helsinki: Ethical Principles for Medical Research Involving Human Subjects, Edinburgh, Scotland, October 2000," *World Medical Association Recommendations Guiding Physicians in Biomedical Research Involving Human Subjects* (Ferney-Voltaire Cedex, France: World Medical Association, 2002).

13. Cf. "The Belmont Report: Ethical Principles and Guidelines for the Protection of Human Subjects of Research" (1979), in Reich, *Encyclopedia of Bioethics*, pp. 2762–2773; and Tom L. Beauchamp and James F. Childress, *Principles of Biomedical Ethics*, 2nd ed. (New York: Oxford University Press, 1983).

14. Conselho Nacional de Saúde, "Resolução N 196/196 de 10 de Outubro de 1996," II.2.

Five

WHO SETS THE AGENDA FOR HEALTH RESEARCH IN DEVELOPING COUNTRIES? A CALL FOR MORE COMMUNITY INVOLVEMENT

Angela Amondi Wasunna

1. Health Research in Developing Countries

Recent international debates on the ethicality of some clinical research trials in developing countries signal the increasing role that poor countries are playing in international health research. The number of new drugs being tested in clinical trials is growing, and the need to find clinical sites capable of conducting human research is growing accordingly.[1] Research sponsors feel pressure to conduct clinical trials expeditiously and inexpensively.[2] Researcher sponsors may be attracted to resource-poor countries where many people with a high disease burden might serve as research volunteers. In those countries, access to patients may be easier due to fewer competing clinical trials and lax regulation. Lower personnel costs may lower the expense of conducting research. The conduct of international biomedical research raises many new ethical challenges including the need for universally accepted ethical principles that can be adapted to the unique cultural norms of each society, the need to address issues related to health inequities between rich and poor, and the need for procedural guidelines to ensure compliance with international ethical standards.

2. Categories of Health Research

Categories of health research include:

(1) Research responsive to a health need of the community that falls within the health research priorities of the community;
(2) Research responsive to a health need of the community does not fall within health research priorities of the community; and
(3) Research not responsive to a health need of the community, and that does coincide with health research priorities of the community.

Since the third category has already been the subject of much international debate, I will not extensively address it in this chapter.

3. The Obligation for Research to be Responsive to Community Health Needs

Because such research increases the risk of exploitation, several international ethics guidelines have endeavored to protect research volunteers in resource-poor countries. The recently revised World Medical Association's Declaration of Helsinki, the 2002 Council for International Organizations of Medical Sciences (CIOMS) *International Ethical Guidelines for Biomedical Research Involving Human Subjects*,[3] and United Nations Joint Programme on AIDS *Guidelines on HIV Vaccine Research*,[4] for example, have strived to create safeguards to protect research volunteers in resource-poor countries from exploitation.

Guideline 3 in the CIOMS *International Ethical Guidelines* states:

> The health authorities of the host country, and a national or local ethical review committee, should ensure that the proposed research is responsive to the health needs and priorities of the host country, and meets the requisite ethical standards.

Guideline 10 also deals with responsiveness of researchers:

> Before undertaking research in a population or community with limited resources, the sponsor and the investigator must make every effort to ensure that: The research is responsive to the health needs and the priorities of the population or community in which it is to be carried out; and any intervention or product developed, or knowledge generated, will be made reasonably available for the benefit of that population or community.[5]

The Declaration of Helsinki states in paragraph 19:

> Medical research is only justified if there is a reasonable likelihood that the populations in which the research is carried out stand to benefit from the results of the research.[6]

The United States National Bioethics Advisory Commission (NBAC), in its report, *Ethical and Policy Issues in International Research: Clinical Trials in Developing Countries*, recommended that "clinical trials conducted in developing countries should be limited to those studies that are responsive to the health needs of the host country.[7]

The British Nuffield Council in its report, *The Ethics of Research Related to Health Care in Developing Countries* also addresses the issue of relevancy of research, explaining:

> [To] enable effective collaboration with external sponsors, developing countries should have a mechanism allowing them to set priorities for research into health care, together with a robust mechanism for scientific review and ethical review of any proposed research.[8]

The ethical obligation for research to be responsive to the health needs of communities in which such research occurs, is well established and has gained consensus around the world.

4. Research Responsive to a Health Need of the Community that Falls within the Health Research Priorities of the Community

Research responsive to a health need of the community that falls within the health research priorities of the community is the most optimal scenario for health research. All researchers, given the limited resources available for health research in developing countries, should strive towards this type of research as their ideal. Most international guidelines require that research meet this standard.

This raises a host of questions and issues. For example, do developing countries have unambiguously stated health research priorities? If so, how have these priorities been determined? If not, how can researchers know whether their research falls within the priorities of the host community or country? Should all externally funded research fall within nationally defined priorities? Do these rules apply to locally funded research? Before responding to these questions, I will address why health research in developing countries should prioritize national and community needs.

A. Bridging the Research Gap

Ninety percent of all health research in the world focuses on diseases that cause only 10 percent of the global burden of disease.[9] Diseases prevalent in poor countries where research is conducted may not be the subject of research at all. For poor countries to bridge this research divide, they must define their health priorities and seek resources to fund research that addresses these identified needs. Internationally, research sponsors and researchers must adopt equity as the core value in setting priorities for research in developing countries.[10] According to the Global Forum for Health Research, the research priorities for developing countries should include child health and nutrition (including diarrhea, pneumonia, HIV, tuberculosis, malaria, other vaccine preventable diseases, and malnutrition); maternal and reproductive health (including mortality, nutrition, sexually transmitted diseases, HIV, and family planning); non-communicable diseases (including cardiovascular diseases, mental illness, and disorders of the nervous system); injuries; and health systems and health policy research.[11] Countries have an ethical obligation to set their health spendingbased on sound scientific, cultural, and economic information. Identified local needs should influence these expenditures.

B. Setting Research Priorities at the National Level

Today, many governments are developing priority-setting processes for health research. In Africa, World Health Organization and the Commission on Health Research for Development (COHRED) recommendations have influenced much of the priority-setting activity. COHRED is an international, non-governmental organization that aims to support, broaden, and strengthen health research linkages and competencies of different stakeholders within countries and at the international level. In 1990, COHRED urged countries to undertake Essential National Health Research (ENHR) to help correct imbalances in global health and development.[12] Under the ENHR plan, priority setting, one of seven elements addressed, should be driven by demand, and focus on an analysis of health needs, expectations, and societal trends. The other elements are promotion and advocacy, ENHR mechanism, capacity building, financing, evaluation, and networking. The involvement of different stakeholders has to be multi-level, with input ranging from the grassroots to the national level, and multidimensional: qualitative, quantitative, and taking into account socio-economic and political considerations.[13]

Over thirty-five developing countries have started experimenting with, or implementing, health research priority setting as a logical component in their essential national health research strategies. These priorities are being developed not only using technocratic approaches, but as part of a comprehensive, participatory, interactive, and iterative process. The input of researchers, decision-makers at different levels, health service providers, and communities are factored into priority-setting decisions. Examples of countries in Africa that are developing research priority plans include Tanzania, Benin, Uganda, Kenya, Ethiopia, Guinea, Zimbabwe, Mali, South Africa, Senegal, Malawi, Sudan, Cameroon, and Burkina Faso. These countries have recognized that prioritization is a political process that involves dialogue and debate as well an underlying value system.

Despite its good intentions, the ENHR Program has its share of problems. Progress in implementing the strategies has been slow due to ineffective communication strategies, weak national funding arrangements, and influences of international organizations involved in health research. Still, many African countries have endeavored to set national and community level health research prioritization plans, and these efforts should be encouraged by researchers and research sponsors.

C. Responsibility

Researchers and research sponsors who conduct health research in developing countries have an ethical obligation to find out what the health research priorities in that country are, to determine whether their proposed research fits within the country's plan. They should ensure that the research plan will not be a duplication of existing research efforts, because all too often identical research projects are being conducted by different research groups in a single country, without any coordination or interaction amongst them. This is a waste of precious resources.

Merely assuming that countries do not have health research priorities is not acceptable. If researchers, local and international, are serious about bridging the research divide, they must identify what institutions are responsible for setting research priorities in that country. In South Africa, Malawi, and Sudan, for instance, the relevant institution is the Ministry of Health, while in Kenya and Tanzania, the governments relegate this duty to their national research institutions. In other countries, for example, Bangladesh, non-governmental agencies have initiated the national research priority-setting process. Researchers should also pay attention to the country's national and community research plans (in development or already developed) and explain how these plans will enhance the national or community agenda. If researchers discover that a country does not have preexisting statements of health research priorities, they should engage with stakeholders in the country, including community members, government officials, local researchers, and local health providers, to determine explore the importance of the proposed research in the local context, and explain how the community might benefit from the results of the study.

D. Motivation

In an NBAC Survey of American Researchers Working in Developing Countries, 73 percent said that their interest in addressing global inequalities in health motivated them to work in a developing country. But 40 percent said that the research priorities of their funding agencies were inconsistent with the top priorities of the developing country in which they were conducting research.[14] Researchers who are genuinely interested in diseases of the developing world need to lobby (in collaboration with other organizations) funding agencies to pay more attention to the 90–10 percent research gap. For this to happen, developing countries must have well-defined health research priorities. The EHNR initiative has been instrumental in this regard, and international researchers should engage to the extent possible with this process to become familiar with the health priorities in the countries in which they propose to work.

5. Research Responsive to a Health Need of the Community that Does Not Fall within Health Research Priorities of the Community

We see problems when research is responsive to an identified health need of a community, but does not fall within the health research priorities of that community. Should we allow the research in these situations? If so, under what conditions? If no, why not?

I do not think that we should exclude all externally funded research that falls outside the nationally defined list of priorities, but we should consider some caveats. The Nuffield Council Report states, for example, that since all research has the potential to contribute to the development of local skills and expertise quite apart from the inherent value of diversity in research; such studies should be allowed to take place. The Council recommends that we ensure care be taken to restrain any tendency on the part of a sponsor to pursue their interests to the detriment of those of the host country or its people. The Council suggests that when research, funded by external sponsors, is proposed, that falls outside of the national priorities, those proposing such research should be required to justify the choice of the research topic to the appropriate Research Ethics Committees (RECs) in the host and sponsoring countries.[15]

I would argue that researchers and research sponsors should go further than merely justifying the choice of research topic. Even though the research topic may not be on the list of research priorities, the research might provide other high-priority benefits to the community. While the research would consume time and resources of the host country, it could also provide employment for trained personnel. To determine what these other benefits might be, researchers should engage with stakeholders in the community to the fullest extent possible—from health officials to policy makers to members of the public.

We can then view the entire research process in a more global sense, as a negotiation process where the community identifies needs external to the research topic that it values and from which the community will benefit. Currently, some developing countries enter into "prior agreements" or "community benefit agreements" that lay out realistic plans for making effective interventions or other research benefits available to the host community after a study is completed. Where the research topic is not a high priority, the host community and research sponsors might come to similar agreements that commit to "derivative benefits," benefits other than those directly expected from the studied intervention.[16]

The benefits may cover a broad range, for example, the community may negotiate for a new health clinic, a series of wells to provide pure water, or a much needed vaccination program. The point being is that the community should be empowered to negotiate conditions under which such research will take place.

I am aware that many communities may not have the power to engage in a fair negotiation process. Wealthy research sponsors can wield a lot of influence at every level. Developing country governments may not have an adequate infrastructure to monitor research. The potential research subjects are often poor and illiterate, causing an imbalance in bargaining power between the host and the sponsors. This vulnerability creates a need to exercise caution in the form of unambiguous caveats in our research proposals.

Other pertinent questions need answers: Who is included in the community? How do we define "benefit"? Should the benefits of health research be health benefits? Are economic benefits legitimate? Should the benefits be directed only to the individuals who participate in the study? Should the community reap some of the benefits? Should benefits go to the country at large? These are not easy issues to resolve. All the stakeholders need to commit to engage with each other to reach well-balanced, fair solutions.

Entering into benefits agreements will require grappling with complex and challenging ethical questions. The process will require collaborative efforts of multiple players, local and international.[17] Such a negotiation process, though difficult, will provide opportunities for interested parties to engage in much needed discussion about what we should consider fair benefits for the community.[18]

6. Revisiting the Three Categories of Research

Community input in the drafting of protocols for research is crucial. Local researchers and stakeholders should be involved in the design of research studies. We should also focus on the involvement of communities in determining what type of research should occur in the first place, and what conditions, if any, we should attached to such research. I have provided three broad categories of health research. The first is the optimal scenario where the interests of the researchers and research sponsors coincide with the interests of the host community, and the research is of priority and benefit to the community. Researchers, local and international, should strive towards this goal to narrow the research gap between poor and rich. Results from such research also stand a better chance of being translated into national policy and action because they fit into the broader health goals of the government.

The second category describes a situation where the research question may be of some relevance to the community, but does not rank high on the list of health research priorities in that community. I do not think we should completely abandon this sort of research as long as research sponsors are willing to engage with the community to determine under what conditions it should take place and what benefits the community stands to gain. The caveat in that scenario would be that the local community must have the upper hand in the negotiations. This scenario presents a difficult challenge.

Situations where research is not responsive to a health need of the community or coincide with health research priorities of the community have been the topic of much debate. Some international ethics guidelines have tried to prevent this type of research due its potential for exploitation. Some communities might be willing to assume the risks of this type of research in exchange for other benefits that they might consider to be satisfactory compensation. This sounds plausible on paper, but in reality, many resource-poor countries may not have the negotiating or bargaining power necessary to negotiate these benefits with rich sponsors. A great danger of exploitation exists, and that potential for exploitation forms the basis for a strong argument against doing any such research at all.

In sum, research protocols should be prepared in host countries, with input from as many local stakeholders as possible to ensure that the protocols coincide with the health goals of the community and country, and then submitted to sponsors for technical advice and financial support. Such a process will require capacity building so that health officials in developing countries, local ethics committees, and civil society organizations will have the expertise to take control of the research agenda in their respective countries. Researchers and research sponsors have an ethical obligation to make serious attempts to engage with the host countries or communities, not just in the designing of protocols, but also in deciding what protocols to develop that will provide maximum benefit to the health and well-being of the host community or country.

ACKNOWLEDGMENTS

The author presented some of the ideas expressed in this paper at a meeting of the Regional Program on Bioethics, Pan American Health Organization, in Santiago, Chile in October 2003. An earlier version of this paper was published as part of the proceedings of the Chile conference in Acta Bioetica. My thanks to Daniel Fitzgerald for discussing the more controversial aspects of this paper with me.

NOTES

1. Office of Inspector General, Department of Health and Human Services USA, "Recruiting Human Subjects: Pressures in Industry Sponsored Clinical Research," OEI-01-97-00195 (2000).

2. See Nancy Kass and Adnan Hyder, "Attitudes and Experiences of U.S. and Developing Country Investigators Regarding U.S. Human Subjects Regulations," National Bioethics Advisory Commission, *Ethical and Policy Issues in International Research: Clinical Trials in Developing Countries, National Bioethics Advisory Commission*, 2 (Bethesda, Md.: United States Department of Commerce, Technology Administration, National Technical Information Service, 2001), pp. 49–50.

3. Council for International Organizations of Medical Sciences (CIOMS), *International Ethical Guidelines for Biomedical Research Involving Human Subjects* (Geneva: CIOMS, 2002).

4. The Joint United Nations Programme on HIV/AIDS (UNAIDS), *Ethical Considerations in HIV Preventive Vaccine Research* (Geneva: UNAIDS Guidance Document, June 1991); and D. Guenter, J. Esparza, and R Macklin, "Ethical Considerations in International HIV Vaccine Trials: Summary of a Consultative Process Con-

ducted by the Joint United Nations Programme on HIV/AIDS (UNAIDS)," *Journal of Medical Ethics*, 26 (2000), pp. 37–43.

5. Council for International Organizations of Medical Sciences (CIOMS), *International Ethical Guidelines for Biomedical Research Involving Human Subjects* (Geneva: CIOMS, 2002).

6. World Medical Association, "Declaration of Helsinki: Ethical Principles for Medical Research Involving Human Subjects, Edinburgh, Scotland, October 2000," in *World Medical Association Recommendations Guiding Physicians in Biomedical Research Involving Human Subjects* (Ferney-Voltaire Cedex, France: World Medical Association, 2002).

7. National Bioethics Advisory Commission, *Ethical and Policy Issues in International Research*, Recommendation 1.3.

8. Nuffield Council on Bioethics, *The Ethics of Research Related to Health Care in Developing Countries* (London: Nuffield Council, 2002), chap. 10, note 10.9.

9. Global Forum for Health Research, *Promoting Research to Improve the Health of Poor People* (Geneva: Global Forum for Health Research, 1999).

10. Chitr Siithi-amorn and Ratana Somrongthong, "Strengthening Health Research Capacity in Developing Countries: A Critical Element for Achieving Health Equity," *British Medical Journal*, 321:7264 (2000), pp. 813–817.

11. Global Forum for Health Research, *The 10/90 Report on Health Research 2000* (Geneva: World Health Organization, 2000).

12. Commission on Health Research and Development, *Health Research: Essential Link to Equity in Development* (New York: Oxford University Press, 1990).

13. Council on Health Research for Development (COHRED), "Priority Setting for Health Research: Lessons from Developing Countries," *Health and Policy Planning*, 14:2 (2000), pp. 130–136.

14. Nancy Kass and Adnan Hyder, "Attitudes and Experiences of U.S. and Developing Country Investigators Regarding United States Human Subjects Regulations," National Bioethics Advisory Commission, *Ethical and Policy Issues in International Research*, pp. 49–50.

15. Nuffield Council on Bioethics, *The Ethics of Research Related to Health Care in Developing Countries*, par. 2.32.

16. Alice K. Paige, "Prior Agreements in International Clinical Trials: Ensuring the Benefits of Research in Developing Countries," *Yale Journal of Health Policy, Law and Ethics*, 3:1 (September 2002), pp. 35–67.

17. Ibid., p. 46.

18. The Participants in the 2001 Conference on Ethical Aspects of Research in Developing Countries, "Moral Standards for Research in Developing Countries: From 'Reasonable Availability' to Fair Benefits," *Hastings Center Report*, 345:3 (2004), pp. 17–27.

Part Two

RESEARCH AND TECHNOLOGY AS ANSWERS AND AS QUESTIONS

Six

ETHICAL ASPECTS IN INTRODUCING GENETICALLY MODIFIED ORGANISMS FOR PUBLIC HEALTH PURPOSES

Darryl R. J. Macer

1. Introduction

Currently, researchers in several locations are working to develop genetically modified (GM) mosquitoes for control of vectors that transmit diseases such as dengue fever, malaria, and Chagas disease. The use of GM insects for control of human disease can be consistent with common ethical norms of society. The approach raises few intrinsic ethical issues. Still, before we release any GM insects, we need to assess potential impact on the environment and on human health. We need to establish universal minimal standard of risk assessment, applicable to disease vectors, because diseases and vectors can cross national and continental borders. Stakeholders should consider ethical issues including community engagement, informed consent, group consent, and relations between researchers and local populations. Each community needs to decide its policy guidance for ethical genetic engineering and to negotiate agreements with neighboring countries.

2. The Ethics of Disease Prevention

Because of the widely held ethical principle that human life is worth saving, we usually see global support of efforts to improve existing and develop new approaches for preventing, diagnosing, treating, and controlling infectious diseases that cause loss of human life.[1] But societies are vastly diversified, exhibiting a wide range of worldviews and social structures. Despite consensus on the overall goal, methods to achieve these goals, including the extent to which risks to human health, damage to the environment and other living organisms, and how economic costs are balanced with the need to invest in other goals, have sparked intense debates.

Well-defined principles basic to resolving ethical dilemmas can help decision makers construct policy decisions that are more informed. The principle that we should love the life given to us (self-love) implies that each person should enjoy autonomy (self-rule) to work independently out how to balance ethical dilemmas and choices.[2] The Universal Declaration of Human Rights of

1948 started with the premise that all human beings possess equal rights and deserve the opportunity to exercise autonomy. Persons who do not enjoy some basic level of health cannot make choices commonly accepted as normal. Poverty also restricts choice for many people, and in areas plagued with high incidence of infectious insect borne diseases, we often find widespread poverty as well.[3]

At present, we see a great inequality between rich and poor nations in the direction and priorities of research, and in the distribution of and access to benefits that might come from this research. Under any ethical theory, the presence of diseases that threaten the lives more than a billion people worldwide provides a compelling impetus for efforts to eradicate the diseases. Around the world, we see wide diversity in the risks that members of different communities face from infectious diseases due to individual genetic variation in resistance to infectious disease agents, in individuals' nutritional state and immediate environment, families' economic situation with respect to providing barriers to vectors and disease, and in access to preventative and therapeutic medicines. Working towards better global equity is a goal that attempts to meliorate the impoverished conditions into which some people are born and to improve their opportunities to thrive. Rawlsian justice holds this as an ethical mandate. John Rawls, in *A Theory of Justice*, argues that efforts should be made to minimize the variation in all social factors because no one knows before they are born into which situation they will be born. Therefore, everyone would wish for equal opportunity and equal exposure to risk. All people should have the opportunity to be born and grow up in an environment free of infectious diseases, if that ideal achievable.[4]

Human beings have used technology in efforts to make their lives easier and better for thousands of years. The ethical principle of beneficence supports the development of science and medicine, and its provision to those who suffer, because we should continue to make life better. Beneficence is based on the belief that all people have an intrinsic motivation to love doing good instead of harm, expressed as compassion.[5] Efforts that work toward the betterment of others have a universal moral mandate.

The ethical principle of nonmaleficence, do no harm, dictates that we should be reasonably cautious about premature use of a technology before potential risks are understood. Recently some proponents have advocated a total precautionary principle for genetic engineering, which means that no technology with any known risk should be attempted.[6] The Cartagena Protocol on Biosafety, an international, legally binding agreement that regulates international movement of living modified organisms (LMOs), advises this extreme caution.[7] Since no human action can be guaranteed to have zero risk, in practice, these principles are used to assess the relative safety of technology and are central to any public health program.[8]

The ethical issues raised by biotechnology are often the same ones that we debated in applied ethics long before we had modern biotechnology.[9] We can view ethics in terms of ecocentric, biocentric, or anthropocentric concerns. Ecocentric concerns, those that value the ecosystem as a whole, find expression in environmental concerns. Reverence for life applies to every member of the whole ecosystem.[10] Biocentric thinking emphasizes the value of individual organisms, whether plant or animal. Anthropocentric thinking focuses on human individuals. Trends have emerged for more ecocentric views to be included in recent legislation, with protection of ecosystems as a primary goal. While isolating distinct issues is useful, separating human nature from social interactions is not realistic because almost all of human life is composed of social activity, involving many relationships between people and the ecosystem. Different ethics apply when human activity, for example agriculture or urbanization, attempts to either dominate nature or harmonize with the environment.

Literature contains a variety of definitions for health, disease, disability, and meaningful human life. Working to alleviate disease and empower individuals to reach their potential are universal goals for the progress of humankind. Authors apply different persons in the same culture and have a wide range of differences between cultures. Working to alleviate disease and empower individuals to reach their potential are universal goals for the progress of humankind that cross all definitions.

We can apply the basic ethical principles of autonomy, justice, beneficence, and nonmaleficence to decision making in the face of a range of bioethical dilemmas found in medical and environmental ethics. Whether further ethical principles can always be derived from these has been debated,[11] but the general consensus is that these four principles are fundamental across a range of cultures.[12] The emphasis on individuals' rights that has been a feature of theories of bioethics in North America is less developing countries. Other ethics theories focus on the community. These theories hold that individual rights or are not well suited to the community structure of some societies.

3. Bioethics and Molecular Entomology

Genetically modified organisms (GMOs) can be used for public health purposes in a variety of ways. Control of disease vectors is an emerging practical possibility. We already have a long history of altering the behavior of vectors so that they cannot transmit pathogens to human beings.[13] Insects have also long been the targets of attention in agriculture and medicine. While few intrinsic ethical concerns about killing insect pests exist, ecocentric approaches to ethics do raise some objections to modification of ecosystem components, and we need to more seriously address those concerns.

People of many cultures have developed biotechnologies as they cohabitate with non-human animal species in the wider biological and social community. Biotechnology means biological science applied especially in genetic engineering and recombinant DNA technology. For the purpose of this paper, I will focus on biotechnology that uses living organisms, especially GMOs, in the production of goods or services. Modern biotechnology is founded on the same tradition of plant and animal breeding that gave rise to agricultural societies. Since the mid1990s, a growing number of countries have sold foods produced from GMOs.[14] Fierce international debates have argued the environmental and human health aspects of genetically modified (GM) foods, but so far, science has demonstrated no harmful effects of GM foods on human health.[15] Some independent organizations, voicing concern about the environmental impact of gene transfer, have submitted reports containing a wealth of useful information about the technology and commentary on relevant ethical issues to several governments for consideration.[16]

Genetic engineering allows genes exchange of genes in a controlled manner between different species. Since its invention in 1974, it has conjured up images of hope and dread. Public opinion is mixed, but the general approach has global support beyond the dreams of its pioneering scientists.

With the emergence of genomic sequencing, we have now mapped the human DNA, dozens of pathogens, and some disease vectors, for example, *Anopheles gambiae*, the mosquito that carries malaria.[17] Molecular entomology, the study of DNA and the proteins it encodes in insects, is emerging as a serious scientific approach to insect control.[18]

The United Nations Development Program (UNDP)/World Bank/WHO Special Programme for Research and Training in Tropical Diseases (TDR) addressed the concept of genetic control of insect vectors at a 1991 meeting on use of genetically modified (GM) mosquitoes to replace disease vectors. TDR's Steering Committee for Molecular Entomology has outlined a three-pronged approach towards developing genetically modified mosquitoes for malaria control, with similar approaches for dengue fever and Chagas' disease.[19]

The paradigmatic genetic research design with GM organisms works out genetic transformation of the insects in the laboratory and then tests their behavior and vector characteristics before releasing them into the environment. First researchers study the host-parasite interaction. Then they develop methods to transform the vector. Human beings are at risk of harm from naturally occurring pathogen-transmitting vectors. The primary purpose of the project to create modified vectors is to reduce these risks to human beings. For these studies to be successful, trials need to be conducted that demonstrate reduced risk compared with naturally occurring vectors before the modified vectors are released. Finally, they look at population ecology and genetics. Except in cases where scientists use sterile insects to replace the whole population to eliminate the species in an area endemic for disease transmission, researchers explore mechanisms that would safely spread the new gene among vectors in the wild, eventually replacing the population of harmful vectors with non-harmful insects.[20]

Releasing GM insects raises questions the safety of the procedure. We need to thoroughly investigate possible effects of introducing GM insects into the environment and on other species. Issues regarding the desirability and morality of humankind changing entire elements of nature are also legitimate areas of ethical reflection.

While interested parties debate whether to use of funds to combat infectious disease using genomics and biotechnology or devote their limited funds to implementing practical measures to curb vectors and pathogens in the field, we find wide agreement that genetic modification of disease vectors will become a major strategy in the future.[21]

4. Intrinsic Ethical Issues of Genetic Engineering

Studies have compared ethical issues inherent in the process of genetic engineering compared to traditional methods of animal and plant breeding. In genetic engineering, we achieve more precise control of genetic transformations and accomplish cross-species DNA transfer that would not ordinarily occur in nature.

DNA transfer does occur between species, sometimes even between kingdoms, in nature. We know, too, that insects are subject to naturally occurring genetic flux. Genetic engineering does not accomplish results that would never occur in nature. Because the DNA changes can be precisely designed with a stated beneficial purpose, a targeted genetic change made through genetic engineering should be safer than naturally occurring fluctuations or mutations because the change occurs under more control and only after careful deliberation of potential consequences. Still, public opinion surveys have shown opposition to cross-species gene transfer.

Does genetic integrity of organisms in the ecosystem have an intrinsic value that humankind ought not change? If we restrict genetic engineering to intra-species changes, our work would mimic the natural ways that organisms evolve changes in genetic structure, eliminating this ethical concern.

Other ethical issues raised address the *telos* (purpose) of an organism. Teleology is the branch of moral philosophy that studies ultimate causes in nature. A teleological explanation describes phenomena by their design, purpose, or final cause. Some philosophers believe that purpose and design, endowed by a Creator, exist in nature. Some philosophers believe that the ability to alter the *telos* of an organism has profound implications, whereas others say that altering the *telos* has little meaning in the context of human modification of nature.[22] If we believe that every organism has a purpose, then the *telos* is an intrinsic characteristic, and genetic engineering alters the *telos* or beingness of an organism. Whether changes and control through genetic engineering are significantly different from changes made by human beings to animals and plants in farming and modern life is debatable. If we consider this issue in a historical context, we see that affluent cultures have controlled nature in significant ways, for example, through irrigation and sanitation projects. In some developing countries, limited resources have limited the extent to which similar control has been feasible. The ethical issue focuses on whether, and to what extent, human beings may legitimately exercise control over nature.[23]

5. Animal Rights Concerns

An ethical concern in the discussion of animal welfare is whether animals suffer or feel pain. Those who believe that insects do not feel pain argue that manipulating insects has no intrinsic wrongness.[24] For those who believe that insects feel pain, even if we consider the idea of making so-called vegemals, animals that do not feel pain, we are still manipulating life for human purposes without considering the interests of the animal.[25] Animal rights proponents' concern is that living organisms should not merely be treated as a means to the ends desired by human beings. More ethical concern has been voiced about modifying higher order animals, especially sentient ones, than about engineering insect vectors.

In addition to concern for so-called intrinsic characteristics (pain, sentience, consciousness), human society has also placed extrinsic value on some animals. For example, some animals, such as the American eagle, are national symbols, and some people express greater concern about harming them than other animals. We also see biodiversity concerns regarding endangered species, some of which have been expressed by the Convention on Biological Diversity.

While perhaps only followers of the Jain religion in India refrain from killing insects that are considered human pests on moral grounds, some others might object to insects for other reasons. We wonder whether manipulating insects to eliminate disease vectors would be more acceptable to persons with these ecocentric worldviews than traditional methods of insect control that attempt to eradicate entire insect populations.

The total number of species affected by genetic modification of vectors would be significantly less than the number of species affected by use of insecticides. Still some argue that we should not modify or eradicate even insects because human beings should not modify the ecosystem. Proponents of this view fail to recognize that all human activity modifies the ecosystem in some way.

6. Consent from Trial Participants

The ethical principle of autonomy dictates that all research participants should give informed consent before receiving any intervention that has a reasonable risk of causing harm.[26] Obtaining individual informed consent is problematic in some developing countries,[27] due to the limited resources and commonly more paternalistic models of health-care delivery. Adequate investment of time and provision of suitable materials should mitigate these problems and us to obtain informed consent from individuals at direct risk, even if the exact cultural interpretation of the informed consent process may vary among countries.[28]

The risks may not just be those that arise directly from the ability of the vector to carry the target pathogen. There could be a negative impact on human health by altering the behavior of blood-feeding insects. In the case of insects that cannot be restricted to a designated geographic area because they fly for example, the concepts of human subject and informed consent need to be extended beyond focus on the individual.

Basic ethical issues pertain to vector collection and field studies. Many such studies have relied on a researcher waiting for a vector to land on a human host, and then capturing it, hoping to do so before the vector transmits the pathogen to the human bait. On ethical grounds, we question whether researchers could design safer field study methods that do not expose human persons to risk.

The approach developed for population genetics studies may be useful where the community and local authorities are involved in the decision-making process. Accepted protocol for informed consent requires researchers to provide information to potential participants. Disseminating information about the plans and progress of the project and obtaining the consent of any person potentially affected by the release of transgenic insects would constitute ethical conduct of research trials, regardless of whether national guidelines mandate these procedures.

People who lack the means to express their preferences might be abused by the lack of individual or community consent for research in anthropology and epidemiology.[29] In some cases, research forewent participatory dialogue with the local community and neglected to inform participants of potential risks of participation in the research. Often researchers did not seek informed consent and did not offer benefit sharing.

If a study involves human beings, oversight by an ethics committee or Institutional Review Board (IRB) is wise. Increasing numbers of countries have codified these requirements, charging researchers with specified legal responsibilities, typically about the conduct of research or clinical practice at the local and national levels. The IRB usually requires that each human subject in a medical trial give informed consent to be involved in the project.

An international consulting committee for TDR has published model operational procedures and ethical guidelines for the establishment of IRBs.[30] These guidelines are not sufficient for the broad question of how to obtain informed consent for a public health intervention involving thousands of persons where the benefits are not yet demonstrated. A project to introduce transgenic insects will need an ethics committee with a broad overview, while regional ethics committees to consider local issues.

To consider the issue at a local level, as required for obtaining appropriate informed consent, a local ethics committee (or IRB if associated with an institution) open to the target communities is essential. Cultural differences dictate differences in the way informed consent should be obtained.[31] The modified Declaration of Helsinki and the draft Council for International Organizations of Medical Sciences guidelines outline the accepted norm in international ethical guidelines.[32] In cases involving bilateral research collaboration, researchers should adopt the most stringent ethical standards of participating countries, but doing so creates problems for populations with lower literacy levels and different common sense social assumptions than the rich sponsoring countries. The ultimate decisions should be made by the local ethics committee based on the international standards.

Privacy issues arise when questionnaires containing personal data are stored. For public health purposes, all information about individuals involved must linked to other data, but to ensure privacy, the data should only be identifiable by a designated person who uses a coding frame entered into a computer not linked to a network.

Some public health interventions have targeted children without obtaining informed consent. Proponents of these procedures have claimed a therapeutic imperative that presumes children are incompetent to give informed consent, but if they had the capability to understand what is good for them, would want to be involved in programs that would assist them to avoid disease. Critics have charged that in these cases, researchers should obtain informed consent from parents or legal guardians before they subject their children to any interventions.

In family settings where interventions might affect anyone in the family, the question arises whether researchers should obtain informed consent from every individual family member. In these matters, researchers should consider local cultural norms. In some cultures, every individual of reproductive age (mature adults) should be consulted. In those cases, family consensus and individual consent is appropriate.

Researchers should obtain agreement and understanding of children in the community through suitable materials. An efficient way to reach parents might be through communicating with children in schools. Since children are at higher risk for many of the diseases under study, they stand to benefit more, and most parents may want to be involved in the trial, because of the potential benefit to their children instead of themselves. But children should not be exposed to direct risk from therapeutic trials unless no alternative exists. In a community involved in a GM vector trial, researchers would not expect any direct risks to the human population. Normal procedures that do not require additional consent from children could be applied.

7. Duty to the Community when Experimental Manipulations Are done on the Environment instead of on Individuals

Because the control population for a study may also be at high risk of contracting the studied disease, recent research ethics debates have questioned whether researchers have an obligation to the local population to use the best available means of disease control whenever they enter an area for a study. Ethically, we might need some other vector reduction measures when making any interventional study in an area. This thinking could lead to the conclusion that a researcher may ethically be compelled also to provide the best available proven alternative to the study population.

This might sound like an ethically compassionate approach, but one that would confound the research design. International guidelines' treatment of placebo-controlled trials, for example, the Declaration of Helsinki, have been based on medical drug studies as the paradigm. This paradigm is not an apt one to apply to genetic research on disease vectors where the treatment involves altering the environment instead of giving a specified drug to an individual. Provision of the proven alternative to the area of study alters the dynamics of the disease so that the results of the vector field trial differ from what the results would have been had no established alternative been provided. Such an approach would render any conclusions we draw about the experimental treatment questionable.

8. Environmental Risks and Public Consensus

A variety potential ecological and health risks are associated with the release of GM organisms. To date, researchers have reported successful environmental risk containment in over 10,000 field trials of GMOs released into the environment in a variety of countries.

We can consider environmental risk from either an anthropocentric or ecocentric approach.[33] Identified risks include the possibility of horizontal transfer of the novel genes to non-target organisms and possible disturbance of insect ecology.[34] Although human beings cannot consent for other organisms to be modified per se, in scenarios where, in addition to effects on direct subjects, research trials can be expected to present potential harm to other non-human members of a biological community, or where interventions in the environment might affect human beings residing in that environment, researchers should seek community consensus on the proposed intervention.

People evince variations across cultures in their beliefs about nature and life.[35] In some cultures, for example, in New Zealand, people have expressed concerns over the need to preserve the native fauna and flora, which many in the Maori community consider wrong to modify. Some people are willing to sacrifice themselves to protect the environment. Another example of this phenomenon is the preservation of sacred groves in India, done now for thousands of years, even during times of severe crisis and human death. In this case, nearly all the people of that community are willing to die instead of damage a part of the environment that they cherish. Such strong convictions often correlate with religious beliefs about spiritual life after physical death.[36]

Potential risks to the agricultural systems of rural communities also require assessment because animal diseases transmitted by vectors pose risks to farming families. In addition, there may also be risks to wild animals in surrounding areas, which in some ecocentric environmental views have more intrinsic rights to be left undisturbed than domesticated animals.[37] This calls for broad ecological understanding of the ecological impact beyond public health.

Since 2001, researchers conducting trials of GM insects have instrumental in developing regulatory systems for oversight of GMOs/LMOs, but most countries in the world still have not established systems for oversight of GM insect field releases.[38]

Although many surveys have assessed public opinion about GMOs, few surveys ask people about their views on introducing GM vectors or pathogens for disease control. Survey results indicate that people consider GM plants less risky than GM microbes, animals, or human beings. GM insects for public health purposes appear to be intermediate in the scale of benefits and risk perception. A Japanese national sample collected in Japan indicated that 33 percent of the respondents thought use of genetic engineering to make mosquitoes unable to be a vector for human diseases like malaria or Japanese encephalitis would be acceptable; only 16 percent said it would not, but 50 percent said they did not know. Another question showed 54 percent of respondents voiced approval for environmental release of mosquitoes that do not transmit human disease, and an approximately equal number supported release of GM disease-resistant crops, with 19 percent disagreeing.[39]

Knowledge is a necessary but not sufficient condition for the acceptance of biotechnology. In surveys of scientists and the public in Japan between 1991 and 2000, for example, well-educated scientists were often as skeptical of biotechnology as the public were. They shared the same types of concerns.[40]

The failure of the government authorities to protect public health in cases such as Mad Cow disease outbreaks has resulted in higher public trust of non-governmental organizations (NGOs) such as environmental groups that often lobby against genetic engineering. In addition, the media sometimes disproportionately reports negative aspects of genetic engineering because these risks of technology appeal to the media and many people.[41] As a result, the late 1990s saw a dramatic drop in public support for biotechnology in every country surveyed. We believe that all stakeholders, even opponents of scientific research, should be included in open dialogue about research priorities. But for meaningful informed decision making, reports of scientific activities should be accurate and reported without bias.

The widespread effects of introducing GM vectors and pathogens into the environment demands that the consent process be modified for a community model instead of on isolated individuals on which drug trial consent paradigms were developed. We need to tailor the process to each community in which we propose interventions.

The question of whether every citizen has to consent to public health interventions is not a new one, but with the current social transition from paternalistic societies to autonomous societies where informed consent and free choice are valued, we see this ethical issue increasingly emphasized.[42]

Any initial trial proposal might be confronted with the "not in my backyard" philosophy. Socially powerful persons can be quite effective at blocking trials they perceive to be risky, or, conversely, at attracting social resources towards themselves and away from weaker persons in the community. Ethical practice requires equity in distribution of risks and benefits. One way to ensure this would be to commit to a local community that, if the trial were successful, the full-scale intervention would include them from the beginning. In this way, populations willing to bear potential risks would receive compensation in the form of early benefits when the full-scale safe and effective control program is implemented.

Finally, before they propose field trials to local communities, researchers have first gained commitment that financial resources to conduct the research will be available, and that that sustainable use of any control tool developed will be affordable to them.

9. Ethics of Technology Choices

Some ethicists have debated whether we should concentrate limited research dollars on developing technologies that rectify problems or on studies to learn how to prevent those problems from occurring. Not all local communities will share the modern scientific worldview that technical healing is desirable or better than other options. If we hope to gain their trust and willing cooperation, we need to be flexibility in which approaches we investigate to eradicate disease. In the past, paternalistic interventions were imposed on citizens. Civil rights movements have empowered people to make these decisions independently.

International debates over the morality of obtaining patents have featured several ethical issues such as the balance between benefit gained from intellectual property innovation rewards versus exclusive licenses for production of producst. Critics have made strong calls against patenting medical innovations because, in some cases, medical products have been priced well beyond the means of many in the population to afford. Laws on intellectual property vary between countries, despite attempts to harmonize these laws among industrialized countries and members of the World Trade Organization (WTO). Some developing countries are not members of the WTO. Often whether a country will join WTO hinges on intellectual property rights (IPR) issues.

Another concern voiced by some scientists and the public is that genetic engineering is somehow unnatural. We need to examine this issue closely for consistency with other expressed values. Presented with the threat of contracting disease or the need to combat disease already contracted, most people have few concerns about using other unnatural remedies such as pesticides and pharmaceuticals. We could question whether opposition based on the practice being unnatural is consistent with other expressed values.

Given that most mosquitoes do not transmit disease to human beings, we could argue that the natural state of a mosquito is one that does not act as a disease vector. We could conclude, therefore, that modifying a mosquito that does transmit diseases into one that does merely restores the mosquito's original nature.

The ethical principle of nonmaleficence, do no harm, should guide our research activities and choices of technologies to develop. Concern over possible safety and environmental risks raised by biotechnology prompted WHO, the United Nations Environment Programme (UNEP), and the United Nations Industrial Development Organization (UNIDO) to identify and study these safety issues. As a result, a UNIDO/UNEP/WHO/Food and Agriculture Organization (FAO) ad hoc Working Group was formed in 1990 to work out practical guidelines through a series of consultations with international experts and scientists from developing countries. In 1991, the UNIDO/UNEP/WHO/FAO Working Group on Biosafety issued the *Voluntary Code of Conduct for the Release of Organisms into the Environment*. UNDP and FAO generally support the development of genetic technology as long as benefits and risks of the organisms are considered. Countries need to strengthen their ability to establish committees to adequately address ethical, social, and scientific concerns.

The Scientists' Working Group on Biosafety of the Edmonds Institute in Washington D.C., United States, recommended that field trials of vectors genetically engineered to reduce disease should be small scale in terms of the area of dispersal of the vector:

> In the case of an anti-malaria or anti-dengue intervention, such a field trial could involve a single village or an isolated cluster of adjacent villages. No large-scale release should be attempted until the effectiveness is shown in the first trial.[43]

While participants in the United Nations system have reached consensus that we should proceed with the prudently selected use of GMOs, some groups within society continue to urge caution. Some countries have political regimes that do not accept GMOs, and in democracies, the attitude of the government depends in large part on a popular vote. Where two or more countries hold opposing opinions, another issue arises. While we should respect national sovereignty, GM vectors may spread beyond national borders.

The Cartagena Protocol on Biosafety was issued in January 2000, and effective September 2003, by the 1999 Conference of the Parties to the Convention on Biological Diversity (CBD) in Cartagena, Columbia. The objective of the Protocol to the CBD is to contribute to the safe transfer, handling, and use of living modified organisms (LMOs), plants, animals, and microbes, that cross international borders, and to avoid adverse effects on the conservation and sustainable use of biodiversity without unnecessarily disrupting world food trade. The Protocol provides countries the opportunity to obtain information before new biotech organisms are imported. It acknowledges each country's right to regulate bio-engineered organisms, subject to existing international obligations. It also creates a framework to help improve the capacity of developing countries to protect biodiversity. The Protocol also recommends risk assessment and risk management once agreement to import biotech organisms is reached, and capacity building in biotechnology research.

Many developing countries do not have the economic or scientific capacity needed to evaluate the products of modern biotechnology.[44] Based on the Cartagena Protocol recommendations, CBD members have established biosafety clearing houses that function as contact points in each member country, whence this information can be obtained without incurring prohibitive monetary costs. We recommend that information about GM vectors should also be included in these clearing houses.

10. Conclusion

A variety of ethical issues are raised in response to the use of GM insects, but the most challenging may be how to ensure valid informed consent procedures for individuals and communities. Each community or society deserves the opportunity to reach consensus on how it values risk. We need a universal minimal standard of risk assessment applicable to disease vectors because diseases cross national and continental borders.

Before field release of transgenic insects, researchers must assess all associated scientific and social issues and develop safety precautions to address potential risks. They should minimize scientific and social risks through careful design of their research, relevant laboratory experience, and careful choice of the dispersal site. Even if no significant risks are recognized before study inception, a procedure for risk assessment should be set up so that new information can be gathered and interpreted. This procedure may involve establishing a specialized ethical review committee under the auspices of an international body such as TDR to offer advice to researchers on the ethics of projects.

Prior environmental, medical, and social studies should form the basis for site selection. Information should be exchanged as broadly as possible with community leaders and members of the local community. Mass media and existing educational institutions are two vehicles for information dissemination. Mechanisms to obtain individual and group consent need to be developed for public health interventions and in trials where researchers make changes to the environment that can affect members of the community.

One way to entice community involvement in field trials is to promise that the community will be first to enjoy any benefits derived from the research. To avoid intellectual property concerns being a barrier to implementing public health measures using GM vectors or pathogens, prior negotiation about intellectual property rights is preferable to confrontation.

Research data should be disseminated accurately and without bias to all stakeholders to benefit from global expertise and to encourage international consensus. We need an ongoing, active process of ethical analysis through a variety of forums. Researchers can allay most concerns by offering better information and education.

Ethically, we should identify core values that guide us as we research how to modify nature in the service of human needs. The ethical principle of beneficence demands action to eliminate hunger and disease. We must balance the need for technological advancement with our duty to protect and preserve the environment for future generations.

NOTES

1. Daryl R. J. Macer, *Ethical, Legal, and Social Issues of Genetically Modified Disease Vectors in Public Health* (Geneva: UNDP/World Bank/WHO Special Program for Research and Training in Tropical Diseases (TDR), 2003).

2. Darryl R. J. Macer, *Bioethics is Love of Life* (Christchurch, New Zealand: Eubios Ethics Institute, 1998).

3. E. Azevedo and E. de Moraes Marcilio Cerqueira, "Decisions in Circumstances of Poverty," *Eubios Journal of Asian and International Bioethics*, 12:4 (2002), pp. 105–107.

4. *John* Rawls, *A Theory of Justice* (Cambridge, Mass.: Belknap Press, 1971).

5. A. Boyd, P. Ratanakul, and A. Deepudong, "Compassion as Common Ground," *Eubios Journal of Asian and International Bioethics*, 8 (1998), pp. 34–37.

6. Mae-Wan Ho, *Genetic engineering: Dream or Nightmare? The Brave New World of Bad Science and Big Business* (Bath, UK: Gateway Books, 1998).

7. Convention on Biological Diversity (CBD), United Nations Environment Programme, Cartagena Protocol on Biosafety (2001–2005), http://www.bio div.org/biosafety/protocol.asp (accessed 31 July 2005). Contact: Secretariat of the Convention on Biological Diversity, 413 Rue Saint-Jacques, Ste. 800, Montreal, Quebec, Canada, H2Y 1N9.

8. D. Callahan and B. Jennings, "Ethics and Public Health: Forging a Strong Relationship," *American Journal of Public Health*, 92 (2002), pp. 169–176.

9. Gary L. Comstock, *Vexing Nature? On the Ethical Case against Agricultural Biotechnology* (Boston: Kluwer Academic Publishers, 2000).

10. Albert Schweitzer, *The Teaching of the Reverence of Life* (London: Peter Owen, 1966).

11. D. L. Weed and R. E. McKeown, "Ethics in Epidemiology and Public Health, 1. Technical Terms," *Journal of Epidemiology and Community Health*, 55 (2002), p. 857.

12. Tom L. Beauchamp, James F. Childress, *Principles of Biomedical Ethics*, 4th ed. (New York: Oxford University Press, 1994); D. F.-C. Tsai, "Ancient Chinese Medical Ethics and the Four Principles of Biomedical Ethics," *Journal of Medical Ethics*, 25 (1999), pp. 315–321.

13. A. Spielman and Michael D'Antonio, *Mosquito: The Story of humankind's Deadliest Foe* (New York: Faber and Faber, 2001).

14. Clive James, *Global Status of Commercialized Transgenic Crops: 2003*, ISAAA Briefs No. 30 (Ithaca, N.Y.: ISAAA, 2003).

15. U. S. Food and Drug Administration Center for Food Safety and Applied Nutrition, Guidance for Industry, Voluntary Labeling Indicating Whether Foods Have or Have Not Been Developed Using Bioengineering, Draft Guidance released for comment (January 2001), http://www.cfsan.fda.gov/~dms/biolabgu.html (accessed 31 July 2005). Contact : Food and Drug Administration, 5630 Fishers Lane, Rm. 1061, Rockville, Md., 20852; specify Docket Number 00D-1598.

16. Great Britain Royal Commission on Environmental Pollution, *The Release of Genetically Engineered Organisms to the Environment: Thirteenth Report* (London: H.M.S.O., 1989); New Zealand Royal Commission on Genetic Modification, *Report and Recommendations 2001: Report of the Royal Commission on Genetic Modification* (Wellington, New Zealand: Royal Commission on Genetic Modification, 2001); and Nuffield Council on Bioethics, *Genetically Modified Crops: The Ethical and Social Issues* (London: Nuffield Council on Bioethics, 1999).

17. R. A. Holt et al. "The Genome Sequence of the Malaria Mosquito Anopheles Gambiae," *Science*, 298 (2002), pp. 129–149; and C. M. Morel, Y. T. Touré, B. Dobrokhotov, and A. M. J. Oduola, "The Mosquito Genome—A Breakthrough for Public Health," *Science*, 298 (2002), p. 79.

18. A. S. Robinson, G. Franz, and P. W. Atkinson, "Insect Transgenesis and its Potential Role in Agriculture and Human Health," *Insect Biochemistry Molecular Biology*, 34 (2004), pp. 113–120.

19. WHO Special Programme for Research and Training in Tropical Diseases (TDR), *Scientific Working* Group *on Insect Disease Vectors and Human Health Geneva, WHO/HQ, 12–16 August 2002*, Document TDR/SWG/VEC/03.1 (Geneva, TDR, 2002). Contact: Steering Committee for Social, Economic and Behavioral Research (SEB), UNDP/World Bank/WHO Special Programme for Research and Training in Tropical Diseases (TDR), World Health Organization, 20, Avenue Appia, CH-1211, Geneva 27, Switzerland.

20. T. W. Scott et al., "The Ecology of Genetically Modified Mosquitoes," *Science*, 298 (2002), pp. 117–179; and B. J. Beaty, "Genetic Manipulation of Vectors: A Potential Novel Approach for Control of Vector-Borne Diseases," *Proceedings of the National Academy of Sciences* (United States), 97 (2000), pp. 10295–10297.

21. C. F. Curtis, "The Case for De-Emphasizing Genomics in Malaria Control," *Science*, 290 (2000), p.1508; and S. L. Hoffman, "Research (Genomics) Is Crucial to Attacking Malaria," *Science*, 290 (2000), p. 1509; and A. A. James, C. M. Morel, C. L. Hoffman, and C. Curtis, "Present and Future Control of Malaria," *Science*, 291 (2001), pp. 435–436.

22. L. Munro, "The Future Animal: Environmental and Animal Welfare Perspectives on the Genetic Engineering of Animals," *Cambridge Quarterly of Health Care Ethics*,10 (2001), pp. 314–324.

23. Michael J. Reiss and Roger Straughan, *Improving Nature? The Science and Ethics of Genetic Engineering* (Cambridge, England: Cambridge University Press, 1996); and Donald Bruce and Ann Bruce, eds., *Engineering Genesis: The Ethics of Genetic Engineering* (London: Earthscan, 1998).

24. Peter A. Singer, *Animal Liberation* (London: Jonathan Cape, 1976).

25. Darryl R. J. Macer. "Uncertainties about 'Painless' Animals," *Bioethics*, 3 (1989), pp. 226–235.

26. George J. Annas, *The Rights of Patients: The Basic ACLU Guide to Patient Rights* (Carbondale: Southern Illinois University Press, 1989).

27. E. O. Ekunwe and R. Kessel, "Informed Consent in the Developing World," *Hastings Center Report*, 14 (1984), pp. 22–24; F. A. Alvarez-Castillo, "Limiting Factors Impacting on Voluntary First Person Informed Consent in the Philippines," *Developing World Bioethics*, 2 (2002), pp. 21–27; and M. Angell, "Investigators' Responsibilities for Human Subjects in Developing Countries," *New England Journal of Medicine*, 337 (2000), pp. 847–849.

28. Nuffield Council on Bioethics, *The Ethics of Clinical Research in Developing Countries* (London: Nuffield Council on Bioethics, 1999).

29. G. A. Fine, "Ten Lies of Ethnography: Moral Dilemmas of Field Research," *Journal of Contemporary Ethnography*, 22 (1993), 267–294; Arthur Kleinman, "Ethics and Experience: An Anthropological Approach to Health Equity," *Public Health, Ethics, and Equity*, eds. Sudhir Anand, Fabienne Peter, and Amartya Sen (Oxford: Oxford University Press, 2004); A. M. Capron, "Protection of Research Subjects: Do Special Rules Apply in Epidemiology?" *Law, Medicine and Health Care*, 19 (1991), pp. 184–190; B. M. Dickens, "Issues in Preparing Ethical Guidelines for Epidemiological Stud-

ies," *Law, Medicine and Health Care*, 19 (1991), pp. 175–183; L. Gostin, "Ethical Principles for the Conduct of Human Subject Research: Population-Based Research and Ethics," *Law, Medicine and Health Care*, 19 (1991), pp. 191–201; H. L. Chee, L. El-Hamamsy, J. Fleming, N. Fujiki, G. Keyeux, B. M. Knoppers, and D. Macer, "Bioethics and Human Population Genetics Research," *Proceedings of the UNESCO International Bioethics Committee Third Session*, vol. 1 (Paris: UNESCO, 1996). pp. 39–63; and WHO Special Programme for Research and Training in Tropical Diseases (TDR), Scientific Working Group on Strategic Social, Economic and Behavioral Research, 31 May–2 June 2000, Document TDR/STR/SEB/ SWG/00.1 (Geneva, TDR, 2000). Contact: Steering Committee for Social, Economic and Behavioral Research (SEB), UNDP/World Bank/WHO Special Programme for Research and Training in Tropical Diseases (TDR), World Health Organization, 20, Avenue Appia, CH-1211, Geneva 27, Switzerland.

30. TDR Scientific Working Group on Strategic Social, Economic, and Behavioural Research, 31 May–2 June 2000, Geneva, TDR document TDR/STR/SEB/SWG/ 00.1, 2000.

31. Alvarez-Castillo, "Limiting Factors Impacting on Voluntary First Person Informed Consent in the Philippines," pp. 21–27; and R. J. Levine, "Informed Consent: Some Challenges to the Universal Validity of the Western Model," *Law, Medicine, and Health Care*, 19 (2001), pp. 207–213.

32. World Medical Association (WMA) Declaration of Helsinki, amended by the 52nd WMA General Assembly, Edinburgh, Scotland, October 2000, Document 17-C, http://www.wma.net/e/policy/b3.htm (accessed 31 July 2005). Contact: WMA The World Medical Association, 13, ch. Du Levant, CIB—Bâtiment A, 01210 Ferney-Voltaire, France; and Council for International Organizations of Medical Sciences (CIOMS), "International Guidelines for Ethical Review of Epidemiological Studies," *Law, Medicine and Health Care*, 19 (2001), pp. 247–258.

33. D. R. J. Macer, *Bioethics for the People by the People* (Christchurch, New Zealand: Eubios Ethics Institute, 1994).

34. A. Gupta, K. Guha, "Tradition and Conservation in Northeastern India," *Eubios Journal of Asian and International Bioethics*, 12:1 (2002), pp. 15–19.

35. K. S. Aultman et al., "Research Ethics: Managing Risks of Arthropod Vector Research," *Science*, 288 (2000), pp. 2321–2322; and Food and Agricultural Organization, *Genetically Modified Organisms: Consumers, Food Safety, and the Environment*, FAO Ethics Series 2, (Rome: FAO, 2001).

36. M. A. Hoy, "Impact of Risk Analyses on Pest-Management Programs Employing Transgenic Arthropods," *Parasitology Today*, 11 (1995), 229–232; and Nuffield Council on Bioethics, *Genetically Modified Crops.*

37. Holmes Rolston III, *Conserving Natural Value* (New York: Columbia University Press, 1994).

38. Pew Initiative on Food and Biotechnology, *Bugs in the System: Issues in the Science and Regulation of GM Insects* (Washington D.C.: Pew Initiative on Food and Biotechnology, 2004), http://pewagbiotech.org/research/bugs/bugs.pdf (accessed 31 July 2005). Contact: Pew Initiative on Food and Biotechnology, 1331 H Street NW, Ste. 900, Washington, D.C., 20005.

39. Masakazu Inaba and Darryl R. J. Macer, "Attitudes to Biotechnology in Japan in 2003," *Eubios Journal of Asian and International Bioethics*, 13:3 (2003), pp. 78–89.

40. Darryl R. J. Macer and Mary Ann Chen Ng, "Changing Attitudes to Biotechnology in Japan," *Nature Biotechnology*, 18 (2000), pp. 945–947.

41. John Durant, ed., *Biotechnology in Public: A Review of Recent Research* (London: Science Museum for The European Federation of Biotechnology, 1996).

42. N. E. Kass, "An Ethics Framework for Public Health," *American Journal of Public Health*, 91 (2001), pp. 1776–1782.

43. Edmonds Institute, The Scientists' Working Group on Biosafety, *Manual for Assessing Ecological and Human Health Effects of Genetically Engineered Organisms* (Washington, D.C.: Edmonds Institute, 1998).

44. K. Chinsembu and T. Kambikambi, "Farmers' Perceptions and Expectations of Genetic Engineering in Zambia," *Biotechnology and Development Monitor*, 47 (2001), pp. 13–14.

Seven

THE ETHICAL REVIEW OF RESEARCH INTO RARE GENETIC DISORDERS

Michael Parker, Richard Ashcroft,
Andrew Wilkie, and Alastair Kent

1. Rare Cases

> A medically qualified researcher was telephoned by a clinical geneticist to discuss a patient with an unusual combination of clinical features, presenting a dilemma for diagnosis and counseling as to the correct genetic risk. Two months later, the clinician sent the researcher DNA from the patient, together with clinical photographs and copies of clinical letters. Investigation of mutation hotspots in relevant genes was negative and the sample was added to a "research panel." Further contact was made by the clinician two years later, enquiring whether there were any positive results (the reply was negative) and providing some further clinical information. Eventually, a further year and a half later and after tests of thirteen genes had proved negative, the researcher contacted the clinician to say that a potentially pathogenic change in the DNA had been identified. A request was made for the clinician to obtain samples from the unaffected parents. The mutation was shown not to be present in either parent, establishing that the change had arisen de novo in the patient and was the cause of their clinical problem. Hence, three and a half years after the initial contact, the research had established unequivocally the correct diagnosis, mechanism of inheritance, and appropriate molecular test for the patient's condition.

Many of the most serious genetic conditions are extremely or moderately rare. Despite their rarity, the implications of such conditions for the individuals and families who suffer from them mean that research into their biological causes and patterns of inheritance is valuable. The rarity of such conditions, combined with the seriousness of their effects, also means that, in many cases, such as in the case study presented above, a close relationship exists between the attempt to achieve a clinical diagnosis and an accurate assessment of risk on the one hand, and research into the biological mechanisms at work on the other. While such research may evolve into a larger study involving a search for affected individuals nationally and internationally, a study will typically begin in clinical practice as a doctor strives to discover the information necessary to help the family.

The close relationship between research on rare genetic disorders and clinical treatment of the same disorders has a tendency to cause problems and uncertainty in research ethics reviews and often leads to delays in the approval of research. This has notable implications for the viability of such research. Often groups raising money for research are small patient and family organizations who find it hard to raise sufficient funds to facilitate the research and to accommodate the significant downtime required to deal with lengthy REC paperwork. The process of research ethics review is disproportionately burdensome for such groups and, when compounded by delays caused by disagreements about the acceptability of the research methods, some projects lose viability. This situation has led some critics to argue that such investigations should not be subject to research ethics review but should be considered clinical investigations instead, subject to the ethical regulations governing that type of activity. In this paper, we explore arguments for and against this proposal and suggest some ways in which the process of ethical review might be made more sympathetic to research on rare inherited disorders.

2. The Problem

One advantage of characterizing clinical investigations of rare genetic disorders as research is that the investigators may be able to access research funding that would otherwise be unavailable to them. For example, funding from research councils and nongovernmental organizations (NGOs) such as the Wellcome Trust might be more available. In addition, the best researchers will be willing to devote their time to the projects, and peer-reviewed journals will be more likely to publish the findings of such research, raising the profile of the rare conditions and the research.

Despite these advantages, the label research means that, in the United Kingdom, such activities require approval by Local RECs (LREC) and Multi-Centre Research Ethics Committees (MREC). The rarity of such conditions, the small number of researchers specializing in the condition, and the research centers being spread around different parts of the country generally results in the requirement that research in these areas requires MREC and LREC approval. These requirements are problematic for those involved in research into rare conditions because their projects tend to be reviewed using criteria more suited to large clinical trials than to the kind of activities described in the case above.

Research into rare inherited conditions falls foul of the current regulatory structure in three ways exemplified in the case above. These are:

(1) RECs require consent from research participants to be "closed," meaning samples must be used for a well-defined purpose adequately understood by the research subject in advance, and the sample cannot be used in any way not explicitly stated in the con-

sent. The small sample sizes available in research on rare conditions mean that the process of identifying potential disease-causing mutations is a type of informed "fishing trip," and the consent given in such cases rarely rises to the standards of informed consent explicit or implicit in the model currently used by RECs.

(2) Research into rare conditions also tends to engender criticism for failure to achieve standards for levels of anonymity and confidentiality.

(3) The UK National Health Service (NHS) Research Governance Framework states:

> We can define research as the attempt to derive generalizable new knowledge by addressing explicitly defined questions with systematic and rigorous methods.[1]

RECs sometimes judge that the methods used in research on rare conditions (such as that described in the case at the start of this paper) fail to meet the required methodological standards.

Faced with these problems, those who carry out or fund such research appear to have two alternative courses of action: They can redefine their activities as clinical investigations, which would enable them to side step the need to engage with the REC regulatory structure. The concomitant disadvantage is that obtaining funding from the usual sponsors of medical research would become more difficult. Alternatively, they can accept the need, in some cases, to subject these activities to research ethics review, but work actively towards the development of a regulatory structure more flexible and sympathetic to rare disease research. This would require engaging actively in educating RECs and would be likely to require the funding of empirical research into the ethical attitudes of patients and patient groups regarding issues such as appropriate standards of confidentiality. We hold that the only realistic option is the second.

For the remainder of this chapter, we will address the questions of when clinical investigation becomes research and the characteristics of what would constitute a more sympathetic approach to the review of research into rare inherited disorders.

3. Research and Clinical Practice

A key step in developing a more appropriate regulatory approach will involve reaching pragmatic agreement about which activities should be consider clinical and which should be considered research. The classification is not easy to make. Much work on rare diseases, especially in the early stages with individuals and their families, is relatively easy to conceptualize as clinical investigation, and in other cases, few would deny that the activity is research. Many activities such as described in the case study presented at the beginning of this paper fall uneasily between the two positions when analyzed in light of current definitions of the terms. For this reason, coming to some agreement about how we classify such activities is essential.

A. Arguments in Favor of the Use of the Term Clinical Investigation

The argument in favor of characterizing our case example as a clinical investigation stems from the primary aim of the investigation, which, in the early stages, before it involves the search for cases outside the family, is not to generate new knowledge, or even to generate interventions for the wider group of patients with this condition. Such cases often arise directly from a routine clinical referral, without direct solicitation on the part of the researcher. The primary aim of the clinician and the researcher in such cases is to provide individuals and their families with diagnoses and some information about inheritance risk.

Some commentators have alleged that one of the ways in which these activities differ from clinical practice is their involvement, from an early stage, of family members other than the patient. This is a misunderstanding of the nature of day-to-day clinical practice in genetics. While the involvement of the patient's wider family may, to those unfamiliar with genetics, look like research, standard clinical practice in genetics often needs to involve several family members, for example to draw up a family history.

Considering this type of activity as clinical practice has the advantage of avoiding the need for regulation by RECs while the standards of consent and confidentiality in such practice would continue to be subject to appropriate levels of regulation. As clinical practice, for the example, the investigation would be subject to the law (for example, on battery and negligence) and to professional guidance (such as that of the United Kingdom's General Medical Council[2]). This means that defining such activities as clinical practice need not imply unacceptably low standards of patient protection.[3]

B. Arguments in Favor of the Use of the Term Research

While our case example has some features of clinical investigation, especially in the early stages, that strongly resemble clinical practice, good reasons also exist for considering this sort of activity to be research.

First, the feature of these activities that makes them look un-rigorous to an REC, that they are clinical "fishing trips," is the same feature that makes them look more like research than clinical practice. Unlike most cases in clinical practice, diagnosis, when eventually made, identifies a previously unknown genetic mutation.

Second, reaching an understanding of the nature of the biological mechanisms at work typically takes years.

Third, peer-reviewed journals commonly publish the results and findings of these investigations. These papers look like research reports and publishers often require that authors confirm that they obtained REC approval prior to beginning the investigation.

Fourth, even though normal clinical practice in genetics often requires the involvement of several family members, their involvement in the kinds of activities under discussion can often evolve into a kind of involvement that everyone would agree was research. The most obvious point at which this happens is when case finding moves from the identification within a single family to a search for additional affected individuals and families either nationally or internationally. This is often characterized to the initially referred family in terms of a search for a greater good—the hope that the experience of the individual being studied will benefit others.

Finally, and most important for the groups seeking to undertake such research, a primary reason for wanting such investigations to be classified as research is that this provides access to sources of funding that would otherwise be unavailable.

C. When Does Clinical Practice Become Research?

Some distinctions between investigations that constitute clinical practice and those that constitute research are easily discernable. An investigation that moves from working with a single family to one that actively solicits affected but unrelated individuals moved into the domain of research. At the other end of the spectrum, when investigators search for genetic mutations upon which to base a diagnosis or attempt to check against samples from blood relatives to see whether this is a *de novo* or inherited mutation, this looks like clinical practice.

Between these two ends of the spectrum, distinctions become less apparent and, we suggest, finding an unambiguous dividing line is unrealistic. We need a pragmatic solution. We suggest that the only feasible solution is to consider as research any investigation that searches for affected individuals, either nationally or internationally, other than the originally identified individual and his or her blood relatives.

4. Toward a More Appropriate Approach to Review of Research into Rare Conditions

The adoption of our criteria for drawing distinctions between research and clinical investigations notwithstanding, the key to success in developing tests and therapies for rare genetic disorders will be the development of a more sympathetic approach by RECs to reviewing research into rare conditions. While a system of research ethics review that did not adequately protect research participants would be unethical, we believe that for the review process to be so burdensome as to undermine the feasibility of good quality ethical research is also unethical.

Some evidence suggests that research into rare genetics disorders under the current system labors under this burden. Another concern is that current regulatory procedures insufficiently consider the views of people directly affected by such conditions.

What would an appropriate process of research ethics review of research into rare genetic disorders look like? Perhaps the best way to start to consider this is in terms of the three core problems described earlier in this paper: consent, confidentiality, and methodology.

A. Open-Ended versus Closed Consent

Obtaining valid informed consent is central to ethical practice in research and clinical settings. Except in exceptional circumstances (for example, research in emergency medicine), research participants should only be included in studies where they have given valid, informed consent before proceeding with any treatment or investigation. What ought to constitute valid consent in research into rare conditions?

Valid consent is voluntary, informed, and competently given. RECs and research ethics guidelines have tended to interpret the requirement for consent to be informed to mean that before consent is given, potential participants must be provided unambiguous, precise, accurate, and closed information. Closed means that consent is only validly informed where participants know beforehand and in detail exactly what is going to happen to them, exactly what is going to happen to any sample/information taken from them, and when the research will be completed. Methodological aspects of research into rare conditions sometimes make providing the participant with a detailed and closed account of these three elements of the research process impossible. Researchers faced with a previously unknown condition may not know exactly what steps they will make beforehand.

This need not imply that ethical, adequately informed consent is impossible in research of the kind described in this paper. If we take special care to provide the research participants with as much detailed information as is known by the researchers, participants may be capable of comprehending the nature of the intervention and the process sufficiently well to enable them to give valid consent. Even the explanation about why a researcher may need to take steps impossible to predict ahead of time would constitute the sort of disclosure that would form the basis for consent.

Instead of knowing in advance exactly what steps will be taken, the true test of validity, we suggest, is whether the participants have sufficient understanding of the research and of their role in it to enable them to make a reasoned and balanced assessment of whether to participate. We believe that given adequate support, participants are capable of understanding research of the kind required for rare diseases. Given the intimate association between those affected and the researchers investigating the condition, consent given by families for the proposed investigation is probably more informed than is the case in many projects where standard protocols are followed.

In any consent process, for participants to provide valid consent is crucial. But in the case of research into rare conditions, any requirement to obtain consent in the manner developed with other sorts of research in mind would halt all investigations. For genetic research into rare conditions to be possible, we should modify the nature of the consent process to reflect the unique, unavoidable aspects and particular circumstances of the proposed research. The use of a standard, inflexible model of consent in all cases will not always be appropriate.

B. Anonymity and Confidentiality

Research into rare conditions, by definition, involves studying unique or nearly unique cases. In these cases, we cannot guarantee the kind of anonymity required by RECs, since drawing a connection between reports of the study and the only persons known to have the condition is easy. As in much research in genetics, researchers need to be aware of family structure and the ways the samples under investigation fit into this structure. This compounds the problem guaranteeing anonymity presents. For families, the issue of confidentiality is largely academic. In the community at large, no-one has ever heard of the condition which affects them, while in the (small) circle of those interested in their condition, everyone already knows who they are!

We agree that high standards of confidentiality and anonymity are essential elements of ethical medical research. When asked, patients, patient groups, and research participants consistently place high value on confidentiality. They also place high value on good quality research and the achievement of results, for example, tests and diagnoses.[4] In rare disorders, these two goals need to be balanced in a way sensitive to the exigencies of such research and to patient and family expectations.

In our view, we should judge the appropriate balance between anonymity and research efficacy according to a combination of empirical evidence about what research participants/patients consider acceptable practice and recognition of the need to protect family members who do not want to be involved in research from family and other pressures to participate.

C. Rigorous Methodology

RECs require researchers to adopt a systematic and rigorous methodology, and fail to approve studies that do not meet the same standards as those required for large trials. But genetic research often begins with a clinical investigation of a single patient who evinces a novel mutation. We contend that research on these unique cases and that with small sample sizes will inevitably require different methods from those traditionally required in large trials.

The regulation and review of research into rare inherited disorders should be focused on ensuring that the research methods are appropriate to the task instead requiring that the methods be identical to those standards with large trials. RECs should recognize that different methodological approaches are appropriate in different kinds of studies. To impose a disproportionate regulatory burden on those wishing to do research into rare genetic disorders is inequitable in that it consumes a greater proportion of the available resources securing REC approval than would be the case in large scale trial that investigate common diseases. For RECs to apply inappropriate standards of rigor to research methodology in their assessment and review of research protocols is unreasonable.

5. Conclusion

Currently, research into rare inherited disorders often encounters difficulties in the process of ethical review. Such problems often appear to stem from the application of inappropriate research models. This sometimes leads to significant delays in the approval of such research—delays that can in some cases threaten the viability of the research and impose an inequitable financial burden on those who fund this research.

In this paper, we argue that the best way to overcome such difficulties is by adopting of our proposed definition of the distinction between clinical practice and research, and the development of a more informed and sympathetic REC process. Patient groups, researchers, and ethicists all have key roles to play in this process. We suggest that we might achieve such changes in four related ways:

(1) By carrying out empirical research into the attitudes of members of families with inherited disorders about key ethical values, for example, on the appropriate balance between respect for confidentiality on the one hand and research success on the other;
(2) By actively educating members of RECs about the appropriate methods of research on rare conditions;
(3) By actively educating and informing research funders and journal editors about the necessity of the close clinical-research interface and about the research methods used in such investigations; and
(4) By adopting a pragmatic approach in which the wishes of those affected by rare genetic disorders are factored into RECs' consideration and given due weight in determining whether to allow such proposals to be given approval.

ACKNOWLEDGMENTS

The authors of this paper would like to acknowledge the contribution of all the patient groups, researchers, clinicians, research sponsors, ethicists, and lawyers who participated in the three workshops on these issues held at St Cross College Oxford, the Medical Research Council in London and the Wellcome Trust Center for Human Genetics in Oxford. Particular thanks to Maggie Ponder, John Gillott and Melissa Winter.

NOTES

1. (UK) Department of Health, *Research Governance Framework for Health and Social Care* (London: Department of Health, March 2001), par. 1.9 http://www.dh.gov.uk/assetRoot/04/01/47/57/04014757.pdf (accessed 31 July 2005). Contact: Department of Health Publications, PO Box 777, London SE1 6XH, UK.

2. (UK) General Medical Council, *Confidentiality: Protecting and Providing Information* (London, General Medical Council, September 2000), http://www.gmcuk.org/standards/archive/secret_v1.htm (accessed 31 July 2005). Contact: General Medical Council, Regent's Place, 350 Euston Road, London NW1 2JN, UK.

3. UK Department of Health, "Good Practice in Consent Implementation Guide: Consent to Examination or Treatment," (London: (UK) Department of Health, 21 November 2001), http://www.dh.gov.uk/assetRoot/04019061.pdf (accessed 31 July 2005). Contact: Department of Health Publications, PO Box 777, London SE1 6XH, UK.

4. Genetic Interest Group (GIG), Minutes from Research Approval and Rare Genetic Disorders, Workshop held 30 April 2003, Oxford, England (Oxford: GIG, 2003), http://www.gig.org.uk/docs/research-approval2.pdf (accessed 31 July 2005). Contact: Genetic Interest Group (GIG), Unit 4D, Leroy House, 436 Essex Road, London, N1 3QP, UK.

Eight

ONGOING AND EMERGING BIOMEDICAL RESEARCH ISSUES AT THE BEGINNING OF LIFE: UNITED STATES PERSPECTIVES

Rosemarie Tong

1. Introduction

Depending on their moral and political values, social and economic condition, and level of scientific development, citizens of different nations develop different concerns about issues surrounding life, especially its beginning, ending, and meaning. Overall, the United States puts more emphasis on the value of each individual's life than most other nations do. Americans tirelessly debate the necessary and sufficient conditions for personhood and struggle to construct unique selves. They diligently write treatises on the question of personal identity and systematically develop technologies that promise to extend the human lifespan.

Many Americans each believe that their individual lives are so precious that the community should make enormous sacrifices to preserve them. During debates about abortion and *Roe v. Wade*, the 1974 United States Supreme Court case that legalized abortion under most circumstances, many Americans identified with the aborted fetus, imagining what it would be for them not to exist. Even though potential but unborn infants cannot know that they are non-existent, many Americans believe that bringing a new life into the world is so precious that women's reproductive rights may and should be abridged do so.

Not all Americans are convinced that pre-embryonic, embryonic, and fetal human life is of the same moral order as human life outside the womb. In April 2004, approximately one million demonstrators marched in Washington, D.C. People from virtually every state in the union came to protest President George W. Bush's reproductive rights policies and to lobby for reproductive rights of women in general. Most of these million marchers would be reluctant to claim that human life, in its early stages, has no moral meaning or value, but they do not believe that human life in its earliest stages is as meaning-filled and rights-protected as human life in its later, more developed stages. For them, when compared, the life of an adult pregnant woman is of greater moral significance than the life of an hours old embryo.

Americans' preoccupation with life and its value accounts for the United States biomedical research community's devotion of so much of its time to debating the ethics of research on fetuses, embryos, and pre-embryos (an embryo that has not been implanted in a woman's womb). In this essay, I focus on a subset of these issues: *in utero* research on fetuses intended for birth; *in utero* research on fetuses intended for abortion; *ex utero* research on living, though non-viable fetuses; using fetal tissue from dead, aborted fetuses to treat living persons; research on pre-embryos destined for implantation; research (especially stem cell research, including therapeutic cloning) on pre-embryos not destined for pre-implantation; reproductive cloning; and ectogenesis. A discussion of these issues may help bioethicists and researchers outside the United States understand why embryonic research they regard as relatively unproblematic is highly contentious in the United States and unlikely to be either morally praised or politically endorsed.

2. The Value of Life and Biomedical Research on Fetuses, Embryos, and Pre-Embryos in the United States

Research on a fetus implanted in a woman's womb involves the fetus's body and the body of the pregnant woman in which it resides. For this reason, *in utero* fetal research may not be conducted absent the informed consent of the pregnant woman. Although some pregnant women willingly consent to *in utero* research on their fetuses, others refuse not only experimental but also established treatments, surgeries, or drugs to correct a defect in the fetus or to prevent a life-threatening disease such as HIV/AIDS in it.

Such nay-saying pregnant women shock many United States' citizens. They believe that pregnant women have a legal and moral obligation to consent to experimental and established medical interventions for the fetus. But attorney George J. Annas disagrees:

> The [already born] child must be treated because parents have an obligation to act in the "best interests" of their children (as defined by child neglect laws), and treatment in no way compromises the bodily integrity of the parents. Fetuses are not independent persons and cannot be treated without invading the mother's body. Treating the fetus against the will of the mother requires us to degrade and dehumanize the mother and treat her as an inert container.[1]

Until the fetus is born, insists Annas, the pregnant woman is legally free either to accept or to reject any treatments for it.

Decision making about whether research on a fetus intended for birth is in the best interests of the pregnant woman and the fetus can be complex. Even more complex is decision making about whether research on fetuses intended for abortion is morally permissible. United States' bioethicists who support non-therapeutic research on fetuses intended for abortion stress that such research is less harmful to them than the fate that otherwise awaits them, death.

In contrast, United States' bioethicists oppose non-therapeutic research on fetuses intended for abortion because these fetuses may not be aborted successfully or even at all. Sometimes fetuses intended for abortion are born alive or pregnant women sometimes reverse their decision to terminate their pregnancies. The possibility exists that, due to the non-therapeutic research conducted on it, a fetus may be born severely defective. Based on these arguments, the prevailing legal and moral view in the United States is that researchers should be prohibited from doing research on fetuses intended for abortion that they would be loathe to perform on fetuses intended for birth.[2]

United States' bioethicists also disagree over whether research on living but non-viable fetuses that are aborted electively or due to miscarriage is permissible. Some claim that such research may be justified depending on the purposes of the research. For example, it may be morally permissible to do research on a non-viable aborted fetus if the knowledge gained could help to save the lives of premature infants. Others disagree with this view. They object to research on a non-viable aborted fetus because it may feel pain, or because, in the case of elective abortion, the mother may permit research on the fetus because she lacks concern for its best interests. Still others believe that because a woman plans to terminate her pregnancy does not mean she is incapable of acting in her fetus's best interests in the event that the procedure results in a live birth. Confronted with the fact that her fetus, though non-viable, is nonetheless alive, the woman will probably not want researchers to cause it pain or to treat it as a mere clump of tissue.[3]

Using fetal tissue from dead aborted fetuses to treat living infants, children, and adults for diseases ranging from diabetes to Parkinson disease is another morally controversial research practice. Some United States' bioethicists worry that pregnant women, who would otherwise not have an abortion, may decide to have one for the good of the human community. Others express the more plausible view that a woman may get pregnant for the expressed purpose of aborting the fetus so that its tissue can be used to treat a loved one, or even her, suffering from a devastating disease. For example, relatively recent report alleged that a woman aborted her fetus so that researchers could use its liver tissue to treat her severe aplastic anemia.[4]

To prevent researchers and the larger society from viewing dead aborted fetuses merely as unwanted clumps of cells that might prove medically useful, many United States' bioethicists recommend that researchers not use fetal tissue from an elective abortion unless the woman would have had the abortion even knowing that tissue is medically useless. They also recommend that no one should approach a pregnant woman about using her dead fetus's tissue until after the abortion decision is made or, ideally, until after the abortion procedure is completed. Only then is it proper for some person, preferably other than the researchers, to seek permission to use the dead fetus's tissue.

Bioethicist LeRoy Walters describes two contexts for research on pre-embryos: research followed by pre-embryo transfer into either the womb of a woman or, conceivably, some sort of artificial womb, and research not followed by pre-embryo transfer.[5] Although some of the ethical questions that arise in research not followed by pre-embryo transfer are quite different from those that characterize research followed by pre-embryo transfer, both types of research necessitate an analysis of the moral status of pre-embryonic life.

In general, we can identify three views about pre-implantation of embryonic human life. The first view is that from the moment of fertilization onwards, a pre-embryo has the same right to life and bodily integrity that any (adult) human being has. No research should be performed on a pre-embryo that would not be performed on an adult human being. At the opposite end of the spectrum is the view that the pre-embryo is merely a cluster of cells. Regarded as such, the pre-embryo has no personal interests and to perform non-therapeutic and therapeutic research on pre-embryos is justifiable.

Between these two extremes is the view proposed by the United States National Institute of Health (NIH) Human Embryo Research Panel in 1994. According to this panel, while not considered a person, the pre-embryo is a form of human life that merits a considerable measure of respect because of the interests that it may one day manifest. In the United States and many other nations, researchers may conduct studies that promise significant human benefits on pre-embryos only before and not after the appearance of the primitive streak of consciousness at fourteen days, a point after which it may be reasonably argued that the pre-embryo has developed one of the minimum necessary conditions for human personhood.[6]

Because the moral status of pre-embryos is just as debatable as the moral status of embryos and fetuses, the ethical issues that emerge in the context of research on pre-embryos produced during the process of *in vitro* fertilization (IVF) for reproductive purposes are similar to the ones that emerge in the context of research on embryos and fetuses. Because the pre-embryo, like the embryo and fetus, cannot effectively express wishes about being the subject of research, a duly recognized surrogate must protect its best interests, carefully weighing the benefits and harms of the research to the pre-embryo and others, respectively. For the most part, United States' bioethicists no longer view IVF as experimental research but as an overall beneficial treatment—as a set of techniques that enable persons who would otherwise not be born to be born. But if IVF researchers announced plans to implant pre-embryos in an artificial womb, or to engage in gene therapy on pre-embryos prior to their implantation in either a natural or artificial womb, chances are that most United States' bioethicists would regard such endeavors as extremely risky research of questionable value to the pre-embryo and to society.

3. The Value of Life and Therapeutic Cloning in the United States

Not all artificially produced pre-embryos are produced are used for reproductive purposes. Some pre-embryos, originally produced for reproductive purposes, are ultimately used for research purposes. In some cases, pre-embryos are deliberately created solely for research purposes.

In November 1998, two groups, one headed by James A. Thomson at the University of Wisconsin, and the other led by John Gearhart at Johns Hopkins University, announced success in culturing human embryonic stem cells. Neither group had any United States government funding. Both groups had funding from the Geron Corporation, a private California biotechnological firm eager to get an exclusive license for commercial uses of the technology.[7] Geron's goal is to be the first United States company to test human embryonic stem cells in human beings.[8] It plans to file for Federal Food and Drug Administration (FDA) approval in 2005 to begin some embryonic stem cell studies on human beings with new spinal cord injuries.[9]

Embryonic stem cells are pluripotent and, in some instances, totipotent primordial cells. They have the potential to develop into *many* (pluri) or even *all* (toti) different types of human cells. Their cell lines are immortal in the sense that they can be cultured indefinitely into a limitless supply of cells. According to many scientists, stem cells will prove useful in treating damaged human cells and tissues (including major organs), testing pharmaceuticals for safety, studying embryo development, and discovering new techniques for gene therapy. According to the Stem Cell Research Foundation (SCRP), stem cell research promises to create new treatments to help millions of Americans: fifty-eight million with heart disease, forty-three million with arthritis, ten million with osteoporosis, over eight million with cancer, four million with Alzheimer's disease, one million with Type 1, and a quarter million with spinal cord injuries.[10]

Although a limited number of stem cells are located in adults and in umbilical cord blood, the best sources of stem cells are the inner mass of a blastocyst (a stage in the development of a pre-embryo that occurs approximately four days after fertilization) and the gonadal tissue of aborted fetuses. That stem cell research is morally controversial is not surprising. Stem cell research requires the prior destruction of pre-embryos or embryos that many people in the United States believe have significant moral status.

Researchers can secure pre-embryos for stem cell research in three ways. They can use so-called spare or excess embryos from IVF clinics. When a couple produces more pre-embryos than is prudent to implant in a woman's womb for fear of causing a multiple pregnancy, for example, clinicians generally advise them to freeze the surplus pre-embryos for possible future use. So advised, the couple is asked to specify their wishes for the frozen surplus pre-embryos should they decide not to use them after all. Would they want their frozen surplus embryos to be discarded, put up for adoption, used for research, or kept frozen for a fee? If the couple opts to have their pre-embryos used for research, then we must ask whether such research is morally permissible. Just because an individual couple views a pre-embryo as their genetic product to do with as they please does not mean that the larger society regards pre-embryos as commodities or property entirely without rights—as "fair game" for research that would not be permitted on adult human persons.

Researchers can advertise for egg donors and sperm donors, and ask them whether they are willing to have their genetic material used for stem cell research. Some United States' bioethicists maintain using pre-embryos deliberately created for research purposes is morally preferable to using pre-embryos originally created for reproduction. They reason that the consenting egg and sperm donors are not emotionally attached to pre-embryos created from their genetic material. Unlike some couples in IVF programs, they will not have second thoughts about the fate to which they have relegated their frozen surplus pre-embryos, death, as opposed, for example, to the possibility of life through the means of embryo adoption.

Other United States' bioethicists disagree. As they view the issue, using deliberately created pre-embryos for research purposes is morally more wrong than using frozen surplus IVF pre-embryos that would otherwise be discarded. They reason that deliberately creating pre-embryos for research purposes may lead to the cheapening and demeaning of procreation and parenting generally and the commodification of all embryonic life.[11] In contrast, using surplus frozen IVF pre-embryos otherwise destined for discard has the advantage of bringing some good out of an otherwise arguably evil act.

Finally, researchers claim that pre-embryos deliberately created by somatic cell transfer (SCNT) instead of standard IVF have the greatest therapeutic potential because from them, cells completely compatible with the patient's cells can be grown. SCNT is a form of cloning. To create a research pre-embryo through SCNT, researchers must fuse the nucleus of one of the patient's somatic cells with an enucleated donor egg cell—an egg cell the nucleus, first removed and then stimulated to fertilize. Although researchers involved in stem cell research maintain that they do not intend to use cloned pre-embryos for reproductive purposes, some United States' bioethicists and policy makers distrust these promises of restraint. As these critics view the issue, we should forsake therapeutic cloning to prevent facilitation of the development of reproductive cloning, which is generally condemned worldwide and in the United States.[12]

In 2001, President George W. Bush attempted to steer a mid-course between those who have no objections to embryonic stem cell research and those who oppose it in all instances. He announced that federal funds may be used for adult stem cell research, umbilical cord blood stem cell research, and embryonic stem cell research on sixty-four (later expanded to seventy-eight) existing stem cell colonies.[13] With respect to the federal funding of existing stem cell colonies, he reasoned that since there was no way to bring back from the dead the already destroyed pre-embryos, some good might as well come from their evil origin. But he dictated that no federal money would be available either to create additional stem cell colonies from spare IVF pre-embryos or to deliberately create "research" pre-embryos.

Although public reaction to President Bush's ruling was generally favorable, his decision disappointed several groups. On one side, some research lobbyists, especially those hoping to find cures for devastating diseases such as Parkinson disease, Amyotrophic Lateral Sclerosis (ALS), and Alzheimer's disease, protested that Bush's decision constituted an unwarranted restriction on potentially life-saving research. They stressed that sixteen or more of the seventy-eight stem cell colonies available for federal research funds had either died or failed to reproduce in laboratory dishes.[14] They also stressed that the owners of the remaining colonies were unlikely to share them with the research community in general.[15]

On the other side, some staunch abortion foes argued that the fruit of evil acts should never be eaten. For example, Wendy Wright, the communications director of Concerned Women for America stated:

> The President's positions contradicts the Nuremberg Code. . . . We should be horrified at the prospect of participating in research on embryos who are deliberately killed for the same reason that we are horrified that gold fillings were taken from the teeth of Holocaust victims.[16]

Significantly, President Bush's ruling only forbade federal funding for research on stem cell lines derived after 9 August 2001. He did not forbid private entities—for example, companies such as Geron—from funding such research. Subsequent to President Bush's ruling, the United States House of Representatives voted, on 27 February 2003, to ban therapeutic cloning and reproductive cloning in an effort to make the process of SCNT illegal in the public and private sectors. HR2505 (3 August 2001), that amends title 18 of the United States Code—the Weldon-Stupak Bill—imposes fines of $1 million or more and prison terms of up to ten years on anyone who performs SCNT. The United States Senate did not pass its version of HR2505 because of the United States public's increasingly favorable attitude toward stem cell research.

Currently, other bills in the United States Senate, which permit SCNT for purposes of therapeutic research but not for reproductive purposes are competing for senators' votes. For example, in 2003, the Human Cloning Ban and Stem Cell Research Protection Act (S303), The Cloning Prohibition Act (S801), and Human Cloning Prohibition (S245, HR 534) were all introduced on the United States Senate floor with bipartisan support. Although no vote was taken on these in 2003, sponsors reintroduced the Human Cloning Ban and Stem Cell Research Protection Act in April 2005.

Not only is the United States federal government entering the stem cell research fray, state governments have also been taking action. For example, in November 2004, California passed Proposition 71 legalized therapeutic cloning and approved State funds of approximately $300 million over ten years beginning in 2005.[17] Despite some later delays in implementation, New Jersey became the second state in the nation, after California, to pass a law legalizing embryonic stem cell research. In April 2004:

> James E. McGreevey, then the governor of New Jersey, signed a bill to establish the nation's first state-supported stem research institute, proclaiming, over the objections of Roman Catholic bishops, that the frontiers of medical science should not be hemmed in by politics.[18]

Established in 1999, the Stem Cell Institute at the University of Minnesota is the world's first interdisciplinary institute dedicated to stem cell research. Currently they are preparing an application to the FDA, and expect to begin adult stem cell trials in early 2006 pending that approval. The institute has already received more than $39 million in NIH funding and more than $43 million in funding overall for stem cell research.[19]

4. The Value of Life and Reproductive Cloning in the United States

Holding the line between permissible therapeutic cloning on the one hand and impermissible reproductive cloning on the other appears to be the general position of United States' society. Most United States' researchers and an increasing number of United States' political authorities appear comfortable with this compromise. They believe that the line they have drawn in the sand of research will continue to hold. The likelihood is that this line, like other boundary lines, will not hold forever.

The siren call of scientific progress combined with the demand of the United States citizenry for reproducing in whatever way they wish, will, sooner or later, embolden researchers to engage in the reproductive cloning of human beings.[20] In contrast to therapeutic cloning, the purpose of which is to secure stem cells for research, the purpose of reproductive cloning is to implant a cloned embryo in the womb of a woman so that she can carry a pregnancy to term. The purpose of reproductive cloning is to secure a particular kind of child—namely, the identical twin of the person whose genetic material was used in the cloning process.

Although United States' bioethicists do not view naturally produced twins as morally problematic, many of them do think deliberately engineered twins are of ethical concern. Critics of reproductive cloning view this technique as intrinsically wrong because it substitutes the process of human replication for human procreation. For centuries, two separate individuals, a male with sperm and a female with eggs, have physically united to co-create a unique person whose chromosomes are a distinctive blend of their two separate sets of chromosomes.[21] In reproductive cloning, two no longer have to become one for procreation to occur. A person can procreate in a solitary manner, as in the hypothetical case of a woman who has one of her eggs enucleated, filled with the DNA from one of her somatic cells, and then inserted into her uterus at the blastocyst stage.

The thought of people replicating themselves in such a solitary manner troubles most United States' bioethicists. They fear that an individual who desires to procreate his or her twin may be rejecting all the good things human beings have traditionally offered each other in the way of community, companionship, and intimacy.

Others disagree with this negative view of reproductive cloning. In their view, replicating human beings is not intrinsically wrong. They argue that like all children, cloned children do have two genetic parents, namely, the two genetic parents of the person who is cloned.[22] For example, if I am cloned for reproductive purposes and decide to get pregnant with my clone, the child to whom I give birth will be my genetic identical twin, and the genetic offspring of my genetic mother and genetic father.

Less debatable than the moral issues regarding cloning is the view that, at present, cloning any mammal is quite unsafe and highly ineffective. To produce Dolly, the world-famous cloned ewe, Ian Wilmut and his colleagues used 277 enucleated eggs. Only twenty-nine of these enucleated eggs developed to the blastocyst stage, and only one of these blastocysts (Dolly) was successfully brought to term.[23] Reports indicate that because an adult somatic cell was used to clone Dolly, her body prematurely aged before she died in 2003.

Until cloning non-human mammals can be shown to be far more safe and efficacious than currently possible, human reproductive cloning should not be attempted. According to experts, producing a single viable human clone would require scores of women to donate eggs and carry embryos, most of which would either not come to term or be born with major deformities.

Although United States' bioethicists agree that much more research on animals is necessary before attempting human reproductive cloning would be considered safe, opponents of the technique admit that safety issues are only one of their concerns. According to John Robertson, an American attorney and bioethicist, among other concerns that have been raised is the fear that the clone's sense of individuality will be threatened, so much so, that the clone will not regard him or herself as a separate and unique person, or that others will fail to view the clone as a distinct individual.

Proponents of human reproductive cloning hold that we have no reason to think that a clone's sense of individuality will be any more threatened than an identical twin's sense of individuality. Presently, most identical twins do not regard themselves as one and the same person just because their genotype is identical. Others, especially parents and family members, typically strive to differentiate one twin from the other. To be sure, exceptions to this rule exist. Some identical twins do try to fuse their separate identities, making a fetish out of dressing alike, for example. Some parents of identical twins do try to provide their twins with precisely the same experiences in a concerted effort to make their twins as identical as possible. In cases where attempts to rear mirror images of the genetic parents become psychologically harmful to their children, psychological counseling may be of assistance, says Robertson.[24] He believes that clone's identity issues will be no worse than those of many "ordinary" people.

Related to the lack-of-individuality objection to human reproductive cloning what Robertson refers to as the lack-of-autonomy objection. We see many variations on the lack-of-autonomy objection to reproductive cloning, but most of them have to do with guardians seeking to get the clone to think, act, and look the same way its genetic source thinks/thought, acts/acted and looks/looked. Instead of letting the clone follow personal preferences and paths in life, the guardians of the clone may be tempted to do everything in their power to get it to follow in the footsteps of the clone's genetic source. [25] They may lock doors that the clone might otherwise have opened, permitting the clone to open only those doors which lead to having the same experiences that the clone's genetic source had.

As Robertson sees it, this objection to human reproductive cloning is overstated. He points out that many parents of naturally produced children often lock doors their children may wish to enter, shoving them through doors the parents select instead. The solution to this problem, whether it arises in the family of a clone or in the family of a naturally-procreated child, is, in Robertson's estimation, psychological counseling.

Another objection to human reproductive cloning that gives many United States' bioethicists pause is that it may lead to the objectification or instrumentalization of children in general. Some people assert that unless they can receive a guarantee that they will have a child whose genotype is precisely the genotype they wish, they do not want any child. Only a child with what they view as the ideal genetic constitution is of interest to them. Any other child would be a great disappointment.

People with this attitude toward children appear to view children as some sort of consumer good. Fortunately, most parents accept the children born to them. If it turns out to be the case that guardians of a clone are significantly more likely than parents of naturally produced children to view the clone not as an end-in-itself but as a mere means to some other ends, then we may have good reason to challenge or even prohibit human reproductive cloning. For example, parents who want to clone an existing child only to use it as a source of spare parts for the first child would be subject to intense moral scrutiny.

An even more powerful objection to human reproductive cloning is the way in which it may undermine kinship, lineage, and family relations. Robertson analyzes the degree to which cloning an existing (or deceased) child, cloning an unrelated third party, rearing a clone of one's self, and cloning and rearing one's parents might threaten the (genetic) family as we know it. As he sees it, kinship issues do not loom at all when genetic parents clone an existing (or deceased) child because they are just as much the (genetic) parents of the clone as of the other child. Kinship issues do emerge, though not dramatically in Robertson's estimation, when the DNA of an unrelated third party is used in the cloning process. He holds that if society is prepared to endorse embryo adoption, it should also be prepared to endorse cloning a third party. If a couple is permitted to adopt embryo *x* instead of embryo *y* because they prefer the genetic characteristics of embryo *x*, why should we not a couple to gestate and rear the clone of a third party whose genetic characteristics they prefer?

Despite his willingness to endorse many types of human reproductive cloning, Robertson does not necessarily accept all forms of human cloning. He suggests that cloning oneself or one's parents may be perceived as too deviant from ordinary reproductive arrangements to fall within the protected sphere of procreative liberty.[26]

5. The Value of Life and Ectogenesis in the United States

Pre-embryos fourteen days old and younger are typically the proposed objects of stem cell research, including therapeutic cloning and reproductive cloning. Research on pre-embryos older than fourteen days, including ones that have been successfully implanted in women's wombs has also been contemplated, though not pursued for technical and ethical reasons. Interest in Ectogenesis—gestating human beings *ex utero* in an artificial placenta—is spurred from the end- and the beginning-side of the gestational process.

In 1984, Peter A. Singer, a philosopher, and Deane Wells, a member of the Australian Parliament, wrote a book entitled *Making Babies: The New Science and Ethics of Conception*, in which they argued that ectogenesis was on the "fast track" for development for two main reasons. First, fetal viability was shifting to ever-earlier times due to developments in the Neonatal Intensive Care Unit (NICU). By the late 1980s, doctors were saving babies born after only six or even five and a half months gestation, weighing fewer than 100 grams. They predicted that before long, someone would develop an incubator able to sustain a five-month or even a four-and-a-half month old human fetus. They described Japanese researchers' success at incubating a partially developed goat kid from 120 days (the equivalent of the twentieth to twenty-fourth gestational week of a human fetus) to full term, seventeen days later. Despite serious developmental problems with the resulting kid (it was unable to stand or breathe independently), the researchers pronounced their experiment a near success.[27]

Singer and Wells' second source of optimism about the development of ectogenesis drew inspiration from the opposite end of pregnancy, the moment of conception instead of birth. They pointed to the then relatively new procedure of IVF. In IVF, eggs are removed from a woman's ovaries and fertilized with sperm outside her body; and at the time Singer and Wells wrote their book, there had already been reports of scientists keeping human embryos alive *in utero* for up to nine days.[28]

In an attempt to decide whether the overall benefits of ectogenesis exceed its overall harms, Singer and Wells discussed what they viewed as the two strongest arguments in its favor. The first of these arguments is that ectogenesis offers an alternative to surrogate motherhood. Instead of contracting a surrogate mother to gestate her and her partner's embryo, a woman who has eggs but is unable to gestate them could instead use an artificial womb to bring the embryo to term.

Ectogenesis would preclude any potential litigation such as the Whitehead/Stern controversy over Baby M., in which Mary Beth Whitehead, a surrogate mother, refused to give the Sterns, the couple who had contracted her services, the baby she had agreed to gestate for them after she gave birth to it.[29] In Singer and Well's estimation, had an artificial womb been available, the Sterns and the Whiteheads could have been spared much grief.[30]

The second strong argument in favor of ectogenesis, says Singer and Wells, is that ectogenesis has the potential to resolve the abortion controversy. Presently, there no way exists to remove a first-trimester or early second-trimester fetus from a woman's body without killing the fetus. But subsequent to the development of an artificial womb, a fetus could be extracted from a woman's body for transfer and further development in the artificial womb. Singer and Wells think such transfers should be recommended because although women have a right to be "rid of the fetus," they do not also have a right to kill it.[31]

In opposition to Singer and Wells' position, philosopher David James argued that even if ectogenesis might be superior to surrogacy arrangements, society should not spend its limited resources developing such a costly technology that would be of benefit to only a small number of people.[32] Philosopher Christine Overall argued that a women's right to an abortion involves far more than a woman's right to have a fetus extracted from her body.

These critics' point is that a woman has the right not to procreate. When a woman seeks an abortion, she probably has fetal extinction and not merely fetal extraction as her goal.

In support of her position, Overall made four claims: First, keeping a fetus alive against the wishes of its biological mother violates that woman's reproductive autonomy. Although we can free a biological mother from her gestational and rearing connections to her fetus by removing her fetus from her, we cannot free her from the genetic connections to her offspring. Since most biological mothers view abortion as a way to erase all their connections to the fetuses, anything less than the death of their fetus will fail to satisfy them.[33] Second, forcing a woman to submit to whatever abortion procedure is most likely to preserve the life of a fetus is "comparable to a compulsory organ 'donation' in which the patient chooses organ removal but does not agree to the subsequent salvaging and use of the organs."[34] Third, because of her close physical relationship to the fetus, "the biological mother is the most appropriate person—perhaps the only one—to decide the disposition of the fetus."[35] Fourth, limiting the biological mother's control over the disposition of the fetus "is yet another example the takeover of reproduction from women."[36]

Philosopher Anne Donchin echoes Overall's fourth concern about depositing women's unwanted fetuses in artificial wombs:

> [I]f extrauterine gestation were to become an established practice, would not many women be pressured to adopt it—"for the good of their baby"? For within the prevailing social framework, once the prac-

> tice was established, it is unlikely that only intentionally aborted fetuses would be nourished in laboratories. Any other fetus considered "at risk" for any reason would count as a potential beneficiary of laboratory observation and intervention.[37]

Donchin's observation is plausible. In the minds of many, ectogenesis would be the perfect solution to the so-called maternal-fetal conflict problem. Instead of seeking to designate crack-cocaine abusing pregnant women as abusive or neglectful, for example, prosecutors would instead seek court orders to remove from the wombs of these wayward mothers their fetuses. Eventually, only women with "safe" uteri might be permitted to gestate their children. We can, as Donchin does, imagine concerned women requesting physicians to transfer their fetus from them into a safer, albeit more artificial, environment.

6. Conclusion

For those who view the forgoing speculations as fanciful, consider science reporter Sharon Begley's comment, "[S]cientists now think that conditions during gestation, ranging from the torrent of hormones that flow from mom to how well the placenta delivers nutrients to the tiny limbs and organs, shape the health of the adult that fetus becomes."[38] She notes that fetal programming has been the subject of two recent National Institutes of Health conferences, both of which sought to establish connections between a variety of diseases and the conditions present in the womb during fetal gestation. These articles report mounting evidence indicating that individuals who have asymmetrical features, such as ears, fingers, and feet, have lower IQs. One theory to explain this correlation is that asymmetrical features may be markers, or indicators, of stress which took place during pregnancy, and that whatever stress caused the asymmetrical features could also have affected the nervous system, causing impairments in the senses, memory, and cognition. Other studies indicate that the genes that affect the so-called stress response may "turn off" if a fetus is exposed to stress hormones within the womb. The growing child may be impaired in his or her ability to handle stress in adulthood. In the words of research physiologist Peter W. Nathanielsz, "The script written on the genes is altered by . . . the environment in the womb."[39]

Perhaps the most significant feature of Begley's article is its conclusion. There Begley recollects the chapter in Aldous Huxley's *Brave New World* wherein he describes how fetuses develop in artificial wombs, the amniotic fluid adjusted with different ingredients contingent on which gender child the parents desire. Pondering this vision, Begley says:

> The quest for the secrets of fetal programming won't yield up such simple recipes. But it is already showing that the seeds of health are planted even before you draw your first breath, and that the nine short months of life in the womb shape your health as long as you live.[40]

The United States' fascination with the beginning of life shows no signs of abetting. Biomedical research on pre-embryonic life is likely to remain politically volatile and morally debatable research in the United States for the near future. Researchers from other nations that do not share United States' views will probably find collaborative stem cell research with United States' scientists problematic. Hopefully, the problems experienced will not be so large as to frustrate attempts to improve the health and well-being of the human community that do not unnecessarily sacrifice present human life to preserve future human life.

NOTES

1. George J. Annas, *The Rights of Patients*, 2nd ed. (Carbondale and Edwardsville: Southern Illinois University Press, 1989).
2. J. C. Polkinghorne, *Review of the Guidance on the Research Use of Fetuses and Fetal Material* (London: Her Majesty's Stationery Office, 1989).
3. Bonnie Steinbock, *Life Before Birth: The Moral and Legal Status of Embryos and Fetuses* (New York: Oxford University Press, 1992).
4. Endre Kelemen, "Recovery from Chronic Idiopathic Bone Marrow Aplasia of a Young Mother after Intravenous Injection of Unprocessed Cells from the Liver (and Yolk Sac) of Her 22 m. CR-length Embryo: A Preliminary Report," *Scandanavian Journal of Haemotology*, 10 (1973), pp. 305–328.
5. *LeRoy* Walters, "Fetal Research,," *Encyclopedia of Bioethics*, ed. Warren T. Reich (New York: MacMillan, 1995), p. 857.
6. (US) National Institute of Health (NIH), *Report of the Human Embryo Research Panel* (Bethesda, Md.: NIH, 1994).
7. Sharon Begley, "Cellular Divide," *Newsweek* (9 July 2001), pp. 25–26.
8. Hollister H. Hovey, "Geron Hopes to Stage Stem cell Tests in Human beings Starting Next Year," *The Wall Street Journal (New York)* (17 March 17, 2004).
9. Ibid.
10. Andis Robeznicks, "States Take Action on Cloning, Embryonic Research," *American Medical News*, 46:24 (30 June 2003), p. 14.
11. John A. Robertson, "Ethics and Policy in Embryonic Stem Cell Research," *Kennedy Institute of Ethics Journal*, 9:2 (1999), p. 125.
12. Charles Krauthammer, "Mounting the Slippery Slope," *Time* (23 July 2001), p. 80.
13. Katharine Q. Seelye, "Bush Backs Federal Funding for Some Stem Cell Research," *New York Times* (10 August 2001).
14. Justin Gillis and Rich Weiss, "NIH: Few Stem Cell Colonies Likely Available for Research of Approved Lines, Many Are Failing," *The Washington Post* (3 March 2004).
15. Ibid.
16. Laurie Goodstein, "Abortion Foes Split over Bush's Plan on Stem Cells," *New York Times* (12 August 2001).

17. Carl T. Hall, "Proposition 71: State Voters Strongly Backing Cell Research," *San Francisco Chronicle* (3 November 2004).
18. Tina Kelley, "In Race toward First Stem Cell Research Institute, New Jersey Stalls," *The New York Times* (31 July 2005).
19. Sarah Youngerman, University of Minnesota Stem Cell Institute, personal communication, 5 September 2005.
20. Maxwell J. Mehlman and Jeffrey R. Botkin, *Access to the Genome: The Challenge to Equality* (Washington D.C.: Georgetown University Press, 1998), p. 99.
21. Testimony of George Annas in *Scientific Discoveries in Cloning: Challenges for Public Policy* (United States Congress Senate Committee on Labor and Human Resources, Subcommittee on Public Health and Safety, 12 March 1997) (347.122/X9), p. 43, available from The Theobold Legislative Library, Legislative Reference Bureau, 100 N. Hamilton Street, Madison, Wisconsin, 53703.
22. John A. Robertson, "Liberty, Identity, and Human Cloning," *Texas Law Review*, 76:6 (May 1998), p. 1410.
23. National Bioethics Advisory Committee, *Cloning Human Beings: Report and Recommendations of the National Bioethics Advisory Commission*, 104, (1997), p. 22.
24. Robertson, "Liberty, Identity, and Human Cloning," p. 1415.
25. Ibid., p. 1416.
26. Ibid., p. 1419.
27. Peter Singer and Deane Wells, *Making Babies: The New Science and Ethics of Conception* (New York: Charles Scribner's Sons, 1985), p. 118.
28. Ibid., p. 117.
29. Phyllis Chesler, *The Sacred Bond: The Legacy of Baby M* (New York: Time Books, 1988).
30. "Excerpts from Decision by New Jersey Supreme Court in the Baby M Case," *New York Times* (4 February 1988).
31. Singer and Wells, *Making Babies*, p. 120.
32. David N. James, "Ectogenesis: A Reply to Singer and Wells," *Bioethics*, 1:1 (1987), pp. 90–95.
33. Christine Overall, *Human Reproduction: Principles, Practices, Policies* (Toronto: Oxford University Press, 1993), p. 67.
34. Ibid.
35. Ibid., p. 68.
36. Ibid., p. 69.
37. Anne Donchin, "The Growing Feminist Debate over the New Reproductive Technologies," *Hypatia*, 4:3 (Fall 1989), p. 144.
38. Sharon Begley, "Shaped by Life in the Womb," *Newsweek*, 134:13 (9 September 1999), p. 51.
39. Ibid.
40. Ibid., p. 57.

Nine

HUMAN STEM CELL RESEARCH AS A PROMISING HOPE FOR HUMANKIND: A CHRISTIAN ETHICAL CONTRIBUTION

Bart Hansen and Paul Schotsmans

1. Introduction

Some significant events settle themselves in the collective memory of humankind where they function for decades as points of reference for future generations. The announcement of the successful cloning of Dolly is such an event. Each of us will remember this thought-provoking occasion because we saw extended media coverage of this breakthrough in medical science. Immediately, world leaders reacted to question how long it would take before the shepherd would be cloned.[1] At the same time, Dolly has narrowed the world's ethicists and theologians' gaze, for by a fixation on her, they have become hyper-focused on one technique—reproductive cloning. More important is the sequel to this medical breakthrough: The successful Dolly experiment led to the possibility of human embryonic and adult stem cell research using cloned cells.[2]

Our contribution focuses on a positive Christian ethical integration of the possibilities of human stem cell research. After a brief clarification of terminology, we will situate the social and legal debate in the European context. We will evaluate the Roman Catholic debate on the moral status of the human embryo, and we will, from our perspective as theologians, reflect on the Christian theological tradition. We will show how ethical discourse on biotechnological issues is shaped by theological themes. Some theists believe that doing embryonic stem cell research usurps the Divine Creator's place. Others believe that this medical breakthrough is in line with previous technological innovations in medicine. We hold a more open and optimistic perspective. We believe that the human being who does stem cell research, as a created co-creator, is not standing up against God, but he is taking full responsibility to actualize God's intentions.

2. A Terminological Clarification

Stem cells are cells that have the remarkable potential to develop into many different cell types in the body. Serving as a sort of repair system for the body, they can theoretically divide without limit to replenish other cells for as long as the person or animal is still alive.

When a stem cell divides, each daughter cell has the potential to either remain a stem cell or become another type of cell with a more specialized function, such as a muscle cell, a red blood cell, or a brain cell. Three classes of stem cells exist: totipotent, multipotent, and pluripotent. A fertilized egg is considered totipotent, meaning that its potential is total; it gives rise to all the different types of cells in the body. Stem cells that can give rise to a small number of different cell types are generally called multipotent. Pluripotent stem cells can give rise to any type of cell in the body except those needed to develop a fetus.

Pluripotent stem cells are isolated from human embryos that are a few days old. Cells from these embryos can be used to create pluripotent stem cell "lines"—cell cultures that can be grown indefinitely in the laboratory. Pluripotent stem cell lines have also been developed from fetal tissue obtained from fetuses that have achieved five to ten weeks development.

Once a stem cell line is established from a cell in the body, that cell line is immortal, no matter how it was derived; each daughter cell cloned will be genetically identical to the one from which it was cloned. The researcher using the line will not have to go through the rigorous procedure necessary to isolate stem cells again. Once established, the researcher can grow identical cells in the laboratory indefinitely. Cells have the capability to be frozen for storage or distribution to other researchers, allowing exact replication of research procedures.

Stem cell lines grown in the lab provide scientists with the opportunity to engineer them for use in transplantation or treatment of diseases. For example, before scientists can use any type of tissue, organ, or cell for transplantation, they must overcome attempts by a patient's immune system to reject the transplant. Using adult stem cells for transplantation, medicine increases human tissue compatibility and immune tolerance, because donor and patient is the same person.

In the future, scientists may be able to modify human stem cell lines in the laboratory by using gene therapy or other techniques to overcome this immune rejection. Scientists might also be able to replace damaged genes or add new genes to stem cells to give them characteristics that can ultimately treat diseases and injuries.[3]

3. Societal Reactions

Following the press release announcing the "Dolly" experiment, many international and regional organizations were quick to condemn reproductive human cloning. Debates over related ethical issues have continued, and several significant publications have recommended caution. The United Nations Educational, Scientific and Cultural Organization's (UNESCO) General Conference adopted the Universal Declaration on the Human Genome and Human Rights.[4] After recalling general principles on the universal protection of human rights and democratic ideals, the text starts with the acknowledgment that human genome research and its applications opens immense opportunities for the improvement of human health. The text is mainly devoted to the limitations and the risks of this kind of scientific research. Article 11 states that "practices opposed to human dignity, such as the reproductive cloning of human beings, are not tolerable."[5] Nations are urged to take measures necessary to prohibit such experiments.

On 15 January 1998, the European Parliament (EP) adopted a resolution that urges members to draft legislation criminalizing human cloning. At the same time, the EP invited members and the European Union to work towards a worldwide ban of cloning of human beings with enforceable sanctions.[6] This resolution reaffirms the position of the EP as expressed in its 1993 and 1997 resolutions.

In 2000, in reaction to the United Kingdom government decision to allow research on embryos created by cell nuclear transfer, the EP adopted another resolution on human cloning. In the 2000 text, they condemned the position of the UK government and reiterated the urgency for members to implement appropriate legislation prohibiting any form of cloning.

The EP's repeated calls for national bans on any type of human cloning suggest failure to recognize a conceptual difference between cloning for reproductive purposes. The EP considers cloning, in any form, to be contrary to morality and the public order. The EP is also reluctant to allow stem cell research or embryonic research on spare embryos, urging researchers to limit the number of embryos they create via *in vitro* fertilization.[7]

Unlike the above-mentioned texts, which, though adopted by respected institutions, have no binding force, the European Convention on Human Rights and Biomedicine and its additional protocols are binding in the ratifying countries, and bear close examination. On 12 January 1998, the Additional Protocol to the Convention of Human Rights and Biomedicine was signed in Paris.[8] This protocol, drawing on the article 18(2) of the Convention, which forbade the creation of embryos for research purposes, banned any form of human cloning (article 1(1)). Countries including France, Luxemburg, The Netherlands, Denmark, Sweden, Finland, Portugal, Spain, Italy, and Greece, signed this protocol.

Whereas, as illustrated by the several documents analyzed, countries have reached consensus on the non-desirability of allowing cloning for reproductive purposes, attitudes towards cloning for therapeutic ends varies and remains controversial. The acceptability of such procedures rests on the promises of current scientific research on embryos and embryonic stem cells. Scientists predict that such embryos and embryonic stem (ES) cells could improve IVF therapy and be invaluable in transplantation and regenerative medicine.

On the international level, under the impetus of France and Germany, the United Nations General Assembly resolved to work towards drafting a binding treaty on cloning.[9] Although an Ad Hoc Committee was created and met as of February 2002, and a draft international convention on the prohibition of all forms of cloning was proposed, the UN decided to delay further discussions on the issue in light of the controversy and lack of consensus regarding the scope of the convention.[10] Whereas some countries, France and Germany included, were in favor of considering reproductive and therapeutic cloning in two separate documents, other nations promoted an all-encompassing document.

Regionally, the EP, despite its earlier prohibitory declarations on all forms of cloning, explicitly rejected the *Fiori Report*, which was a move to ban all forms of human therapeutic cloning in the European Union. On 29 November 2001, only thirty-seven of the 391 Members of the European Parliament voted in favor of the report. Similar voices could be heard at the conference of the Life Sciences High Level Group of the European Commission, entitled *Stem Cells: Therapies for the Future*, held on 19 December 2001.[11] The opinion of the European Group on Ethics in Science and New Technologies, addressed to the European Commission, proposed to allow stem cell research on extra embryos created with IVF that would otherwise be discarded, but considered therapeutic cloning to be premature.[12]

A few countries have, on a limited basis, explicitly authorized therapeutic cloning or remained silent, giving implicit approval. Other countries, prohibiting reproductive and therapeutic cloning, have endorsed the use of spare embryos.

In February 2002, a select committee of the United Kingdom House of Lords granted United Kingdom scientists permission to create and conduct research on human embryo clones. Licenses for the use of cloned embryos are awarded on an exceptional basis, the presence of a demonstrable justification that the research contemplated cannot be carried out by using spare embryos after an IVF-treatment.[13] Promoting stem cell research and demonstrating United Kingdom's scientific leadership, the United Kingdom Government has provided its support to the establishment of the first stem cell bank, which opened on 29 May 2004.[14]

The recent Belgian Embryo Law accepts all types of research directed at therapeutic purposes and increased medical knowledge. This law implicitly includes therapeutic cloning and the development of ES cell lines.[15]

Several other countries of mainland Europe have also taken position on stem cell research and therapeutic cloning. France and The Netherlands have recently adopted a more moderate position on ES cell research. They oppose the creation of embryos by means of nuclear transplantation solely for research purposes. On the other hand, embryos derived from IVF, no longer required for a parental project, can be used for research.[16]

Germany's National Ethics Committee circumvented the rigorous "embryo-law" by recommending importation of human stem cells from abroad.[17]

In the United States, therapeutic cloning remains a controversial issue. The current debate mainly focuses on the question of federal funding. On 9 August 2001, United States President George W. Bush limited funding to research on stem cell lines derived prior to his announcement.[18] For the time being, only research on sixty-four stem cell lines (later increased to seventy-eight) originating from supernumerary embryos qualifies for federal funding.

These international, regional, and national positions, characterized most notably by absence of international consensus on therapeutic cloning, underlie many vigorous ethical debates.

4. The Ethical Debate on the Moral Status of the Human Embryo

Faust, in the legend, traded his soul to the devil in exchange for knowledge. To strike a Faustian bargain is to be willing to sacrifice anything to satisfy a limitless desire for knowledge or power. We can compare the ethical debate surrounding human embryonic stem cell research to a Faustian bargain: proponents of stem cell research are willing to sacrifice human embryos to develop cures for lethal diseases or treatments for devastating injuries such as spinal cord injuries. Can we turn a blind eye on destroying embryos for the sake of stem cell research? To answer this question we need to examine the moral status of the embryo.

The nature and the moral status of the human embryo is a vast and difficult subject in general ethics and in Roman Catholic moral theology. Citing the inviolable integrity of a human embryo from the moment of conception, the Congregation for the Doctrine of the Faith has repeatedly condemned creating human life for research purposes.[19] Despite the position of the Church being univocally condemnatory, debate on the moral status of human embryos continues within the Roman Catholic community. Not all Catholics agree with the official position of their Church. As early 1970s, Karl Rahner mentioned "the uncertain rights of a human being whose very existence is in doubt."[20]

Recently, a growing number of Catholic theologians, including Richard A. McCormick in the United States, Patrick Verspieren in France, and John Mahoney in the United Kingdom, does not consider the human embryo to be an individualized human entity, separate from the mother, enjoying the inherent potential to become a human being. They accept a distinction between genetic individualization, meaning that an individual is genetically unique, present from the beginning of life, and developmental individualization, wherein an individual begins to be a unique person only after a period of physiological development has occurred. McCormick claims, "Developmental individualization is completed only when implantation has been completed, a period of time whose outside time limits are around fourteen days."[21]

Those who make the genetic/developmental individualization distinction appear to prefer a return to the centuries-old Catholic position that a designated amount of development is necessary for a fetus to warrant personal status. Embryological studies confirm that fertilization is a process, not a moment. Some ethicists opine that in its earliest stages—including the blastocyst stage, at which the inner cell mass is isolated to derive stem cells for research—the embryo is not sufficiently individualized to bear the moral weight of personhood.[22]

The more liberal religious view on the status of early human life supports a more open-minded attitude towards research on embryos. McCormick, Verspieren, and Mahoney, while opposing the creation of human embryos solely for research purposes, do not condemn the use of spare embryos from an infertility treatment. These embryos, which would otherwise be discarded, may instead be donated for research.[23] This viewpoint has introduced diversity into Catholic debates about *in vitro* fertilization, pre-implantation genetic diagnosis, human gene therapy, and human embryonic stem cell research.

Although we may question whether a fundamental moral distinction exists between using supernumerary embryos in research and the creation of embryos for research purposes, techniques like IVF would not have been possible without research on embryos solely created to enhance the success rate of such infertility treatments.[24]

5. Human beings as Created Co-Creators

James Walter has described two different theological models of God in relation to creation and humankind present in Christian tradition: the Stewardship model and the Created Co-Creator Model. Each model has different implications for how human biotechnological applications should be judged.[25]

A. The Traditional Stewardship Model: Man as Steward of God's Creation

According to the tenants of the Stewardship Model, God is the creator of the material universe, including humanity and fixed universal laws of creation. Thereby, all elements of nature are sacred, having divine purpose, deserving respect, and not to be challenged or changed. We can understand God's purposes for humanity by scrutinizing the universal laws governing nature.

As a sovereign ruler over the created order, God decides of the future by divine providence. As a lord over life and death, God has rights over creation, which human beings do not enjoy. When human beings take God's will into their hands, they usurp divine authority.

Humanity is the steward of God's creation. Humankind is charged with a moral responsibility to protect and maintain divine creation and order. Being only trustees of creation, God does not allow human beings to alter his original creative will. According to Walter, the Stewardship Model is sacral/symbiotic, meaning that humankind should recognize its sacred symbiotic relationship with nature without attempting to interfere with life's natural course.[26] Human beings' moral duty is to live within the boundaries established by God at the first moment of creation.

Embryonic stem cell research aims to alter the nature of human beings by changing the process of procreation and by changing the genetic structure of cells. According to the Stewardship Model, we should condemn such research, even for the therapeutic purpose of curing lethal diseases. Proponents of this model believe that such research constitutes an expression of human arrogance: by manipulating divine creation, human beings usurp a role that God has not delegated to them as stewards. Such research violates the static and inherent laws of nature: As human inner nature is sacrosanct and inviolable, human beings who attempt to alter the nature of human beings are disobeying God's natural law, immorally crossing divine boundaries.

Principles of evolution are not new. From the beginning of scientific investigation, we have seen that principles we once believed to be static laws of nature are, instead, mutable. All objects of creation are naturally in a state of flux, subject to change by forces intrinsic to the organism (for example, spontaneous genetic mutation) or external to the organism (for example, by some condition that alters the placental environment).

Instead of fixed laws of nature, we now view change and development as the norm. Walter continues:

> The relation between nature and human freedom appears as a dialogue that dynamically evolves over time. It is within that dialogue that human beings learn how to use responsibly material reality as the medium of their own creative self-expression.[27]

Given humankind's current comprehension of science, the Stewardship Model no longer appears to accurately reflect the true nature of humankind within nature. Even if we continue to believe that God created humankind to be steward of creation, we know that creation is not static by its nature. Humankind could not succeed in maintaining some static natural state. In these terms, human interventions that change the nature of organisms does not appear to be as immoral as when judged by the traditional Stewardship Model. In short, we evaluate this model to be unsatisfactory to provide us guidance regarding the morality of stem cell research.

Before we make our judgment about the morality of stem cell research, let us conduct some further analysis of the concept of creation and the omnipotence of God.

B. Rethinking the Traditional Theological Interpretation: The Human as Created Co-Creator

The Created Co-Creator Model offers us an interesting perspective on the relationship between God and humanity, a relationship based on respect, common will, and interactivity. Protestant theologian Philip Hefner introduced the term "created co-creator" in "The Evolution of the Created Co-Creator."[28] Catholic theologian Karl Rahner had, in the early 1970s, anticipated a similar vision of humanity in relation to God and creation.[29] We now propose unraveling the several components of this expression and evaluating each of them.

The term co-creator suggests that creation, or nature, is not a static order brought into existence through a single act of God, but a work in progress, in which humankind can play a role. This idea constitutes the doctrine of uninterrupted creation, or "*creatio continua*," of God.[30] New insights gained from genetics support this idea that creation is an evolutionary process in which God remains continuously active. Every living cell contains DNA molecules. Spontaneous genetic mutations underlie human evolution, ensuring survival of the most robust organisms over time, and faults in the mechanism of cell division that can lead to fatal illnesses such as cancer. This process is the physical manifestation of God's ubiquity and continual influence of creation at even the level of the smallest building blocks of life.

"Co-creator" suggests that the human being stands on an equal footing with God in the creative process. Our ever-growing scientific knowledge, especially in the field of genetics, supports this view. Revolutionary genetic developments have redirected theological anthropology, the study of humankind from a theological point of view, to recognize the evolutional principles at work in nature and humankind's ability to influence those changes.

Thanks to the development of ES cell research, human beings have gained insight into one of the secrets of God's creation. With the aid of cloning techniques, they are able to apply this knowledge to the creation of human tissues (cloning). For example, armed with knowledge of the origins of a genetic disease and the technology to perform genetic transformations, the human being is no longer a passive subordinate of God's creation but may take an active creative role in correcting the genetic disorder.

But possessing the knowledge does not ensure that the humankind will always use this knowledge for the greater good of humanity or the betterment of God's creation. Some people believe that reproductive cloning would not be a moral application of the knowledge.[31]

In opposition to God's intentions, presumed to be inherently good, humans can direct their creativity to either good or bad uses. Scientists may apply their technological ingenuity to reduce human suffering, but may also cause such suffering. Human beings are not divine, because they are only an image of God, and their creations bear, therefore, an ambivalent character of good or bad intentions and right or wrong actions. The ensuring component of this faith is that the human being may consider her/himself as a being in co-operation with God to fulfill the dynamics of creation.

To this notion of "co-creator," Christians believe that they must add the complement, "created." God's work of creation is always distinct from that of humans. While the human beings only dispose of the natural properties of already available matter, God may create from nothing (*creatio ex nihilo*). It implies at the same time that the human being has always in faith to find out what God intends with his creation.

Applying the precepts of this model to form a theological opinion on the use of the new cloning technologies for therapeutic purposes, we should bear in mind two essential consequences of the "created co-created" concept. On the one hand, the nature of humankind, created by God, is good. God has created humans in his image (naturally good). He entrusts them with the freedom and the responsibility to apply technology in such a way that it moves forward the promises of his unending Kingdom (cf. the created *co-creator*). Hand in hand with God, as his partners, human beings may believe that they have received the responsibility to improve the sketch of Creation. On the other hand, believers also are strongly aware of the fact they are not perfect: the human is but a "poor" and "imperfect" image of God, he is a created image not enjoying the same status as the model (as compared with the *created* co-created). They continuously have to search how to realize more good than evil, being aware how weak they sometimes may be.

In faith, we cannot separate these components. Christians believe in a God who makes himself dependent on how human beings try to fulfill the great expectations He has of them. God gave the world in human hands, but is also in his unending Love dependent on how human beings realize in responsibility the great future He has for them in his Spirit. Concerning stem cell research, this implies that human beings have always to evaluate in a proportionate way the benefits and the risks of this promising technology.

6. Conclusion

An ethical evaluation has the challenge to inquire how we can realize the growing possibilities of human technology to serve the dignity of humankind in moral ways. An ethical weighing of values and an open philosophical and theological perspective on human beings as created co-creators leads to a positive evaluation of human stem cell research. The promotion of every human being becomes a moral obligation insofar as possible. Ethics guides living and should evolve as human life unfolds throughout history. This is a continuous, but creative challenge: As a created co-creator, each human being may assume her or his responsibility to take creation one step closer to completion in line with Gods intentions.

NOTES

1. Guido De Wert, *Het Schaap kan nu Worden Gekopieerd, de Volgende Stap is de Gekloonde Herder* (A Sheep Can Be Copied, the Next Step Will Be a Cloned Shepherd), *NRC Handelsblad* (12 March 1997), p. 7.

2. Ronald Cole-Turner, "The Era of Biological Control," *Beyond Cloning: Religion and the Remaking of Humanity* (Harrisburg: Penn.: Trinity Press International, 2001), pp. 1–2; and Ian Wilmut, Keith Campbell, and Colin Tudge, *The Second Creation: Dolly and the Age of Biological Control* (New York: Farrar, Straus and Giroux, 2000).

3. Department of Health and Human Services, *Stem Cells: Scientific Progress and Future Research Directions* (Washington, D.C.: Department of Health and Human Services, June 2001). http://stemcells.nih.gov/info/scireport (accessed 15 August 2005).

4. UNESCO, Universal Declaration on the Human Genome and Human Rights (Paris: UNESCO,1997), http://portal.unesco.org/en//ev.php-URL_ID=13177&URL_SECTION=201.html (accessed 1 August 2005). Contact: UNESCO, 7, place de Fonenoy, 75352 Paris 07 SP, France.

5. Ibid., art. 11.

6. European Parliament, "Resolution on Human Cloning," *Official Journal of the European Communities*, C34 (15 January 1998), p. 164.

7. European Parliament, "Resolution on the Cloning of the Human Embryo," *Official Journal of the European Communities*, 315 (28 October 1993), p. 224; European Parliament, "Resolution on Cloning Animals and Human Beings," *Official Journal of the European Communities*, 115 (12 March 1997), p. 92; and European Parliament, "Resolution on Human Cloning," *Official Journal of the European Communities*, 315 (7 September 2000). Contact: OJEC, 4 Albert Street, Aberdeen, AB25 1XQ, UK.

8. Council of Europe, *Additional Protocol to the Convention for the Protection of Human Rights and Dignity of the Human Being with Regard to the Application of Biology and Medicine, on the Prohibition of Cloning Human Beings* (Strasbourg, Austria: Conseil de l'Europe, 1998).

9. United Nations Resolution General Assembly, resolution 56/93 of 12 December 2001. See Report of the Ad Hoc Committee on an International Convention against the Reproductive Cloning of Human Beings (New York: United Nations, 2002).

10. A Working Group of the Sixth Committee (29 September to 3 October 2003) of the United Nations Ad Hoc Committee on an International Convention against the Reproductive Cloning of Human Beings considered a proposal by Costa Rica for a draft international convention on the prohibition of all forms of human cloning (A/58/73).

11. Life Sciences High Level Group of the European Commission, *Stem Cells: Therapies for the Future?* (Luxembourg, Belgium: Office for Official Publications of the European Communities, 2002).

12. European Group on Ethics in Science and New Technologies, *Adoption of an Opinion on Ethical Aspects of Human Stem Cell Research and Use = Adoption d'un avis sur les Aspects Éthiques de la Recherche sur les Cellules Souches Humaines et leur Utilisation* (Brussels, Belgium: European Group on Ethics in Science and New Technologies to the European Commission, 2001).

13. Great Britain, Parliament, House of Lords, Select Committee on Stem Cell Research, *Stem Cell Research—Report from the Select Committee* (London: Stationery Office, 2002).

14. "World's First Stem Cell Bank Opens in the UK," *BioNews* (20 May 2004) http://www.bionews.org.uk/new.lasso?storyid=2096 (accessed 2 August 2005).

15. "*Wet van 11 Mei 2003 Betreffende het Onderzoek op Embryo's in Vitro*," *Belgisch Staatsblad* (28 mei 2003) (Law of 11 May 2003 Regarding the Research on Embryos in Vitro, Bulletin of Acts Orders and Decrees (28 May 2003); and "*Loi de 11 Mai 2003 Relative à la Recherche sur les Embryons in Vitro*," (Law of 11 May 2003 Relating to Research on *in Vitro* Embryos) (Law of *Moniteur Belge* (28 Mai 2003).

16. Elizabeth Guigou and Lionel Jospin, *Projet de Loi Relatif à la Bioéthique* (*Law Project Relative to Biotechnology*) (Paris: Assemblée Nationale, 2001); and T. A. Braake, "*Wet van 20 Juni 2002, Houdende Regels Inzake Handelingen met Geslachtscellen en Embryo's*," see T. A. Braake, "The Dutch 2002 Embryos Act and the Convention on Human Rights and Biomedicine: Some Issues," *European Journal of Health Law*, 1 (2004).

17. "Gesetz vom 28.*Juni* 2002 zur Sicherstellung des Embryonenschutzes im Zusammenhang mit Einfuhr und Verwendung menschlicher embryonaler Stammzellen," Bundesgestzblatt Jahrgang, ("Law of 28 June 2002 on the Protection of Embryos Linked to the Import and Use of Human Embryonic Stem Cells"), 1:42 (29 *Juni* 2002), p. 2277.

18. George W. Bush, "Remarks by the President on Stem Cell Research," White House Press Release (9 August 2001) http://www.whitehouse.gov/news/Releases/2001 08/20010809-2.html (accessed 2 August 2005).

19. Congregation for the Doctrine of the Faith, *Donum Vitae, Instruction on Respect for Human Life in Its Origin and on the Dignity of Procreation: Replies to Certain Questions of the Day* (London: Catholic Truth Society, 1987); and Congregation for the Doctrine of the Faith (John Paul II), *Encyclical Letter Evangelium Vitae* (Washington, D.C.: Unitied States Catholic Conference, 1995).

20. Karl Rahner, "The Problem of Genetic Manipulation," *Theological Investigations*, 9 (1972), p. 246.

21. Richard A. McCormick, *The Critical Calling: Reflections on Moral Dilemmas Since Vatican II* (Washington D.C.: Georgetown University Press, 1989), pp. 343–346.

22. Margaret A. Farley, "Roman Catholic Views on Research Involving Human Embryonic Stem Cells," in Suzanne Holland, Karen Lebacqz, and Laurie Zoloth, eds., *The Human Embryonic Stem Cell Debate: Science, Ethics and Public Policy* (Cambridge, Mass.: MIT Press, 2001), p. 117; Thomas A. Shannon and James J. Walter, "Reflections on the Moral Status of the Preembryo," *Theological Studies*, 51 (1990), pp. 603–626; and Lisa S. Cahill, "The Embryo and the Fetus: New Moral Contexts," *Theological Studies*, 54 (1993), pp. 124–142.

23. Cf. Gene H. Outka, "The Ethics of Human Stem Cell Research," *Kennedy Institute of Ethics Journal*, 12 (2002), pp. 175–213.

24. Carol A. Tauer, "Responsibility and Regulation: Reproductive Technologies, Cloning, and Embryo Research," *Cloning and the Future of Human Embryo Research*, ed. Paul Lauritzen (New York: Oxford University Press, 2001), p. 153.

25. James J. Walter, "Human Gene Transfer: Some Theological Contributions to the Ethical Debate," *Linacre Quarterly*, 68 (2001), pp. 319–334, esp. pp. 324–327; and Thomas A. Shannon and James J. Walter, *The New Genetic Medicine: Theological and Ethical Reflections* (Lanham: Rowman & Littlefield Publishers, 2003).

26. Walter, "Human Gene Transfer," p. 326.

27. Ibid., pp. 326–327.

28. Philip Hefner, "The Evolution of the Created Co-Creator," *Cosmos and Creation: Theology and Science in Consonance*, ed. Ted Peters (Nashville: Abingdon Press, 1989), pp. 211–233; and Walter, "Human Gene Transfer," p. 332 n. 25.

29. Karl Rahner, "The Experiment with Man: Theological Observations on Man's Self-Manipulation," *Theological Investigations*, 9 (1972), pp. 205–224; and Rahner, "The Problem of Genetic Manipulation," pp. 225–249.

30. Ronald Cole-Turner, *The New Genesis: Theology and the Genetic Revolution* (Louisville, Westminster John Knox Press, 1993), pp. 98–99.

31. Ibid., pp. 102–103.

Ten

INTERRELATIONS BETWEEN BIOETHICS AND ETHICS OF BIOTECHNOLOGY

Boris G. Yudin

1. Two Fields of Ethical Concern

To discuss interrelations of bioethics with ethics of biotechnology we need first to draw distinction between these two fields. Making the distinction is difficult because no generally accepted understanding of either field of ethical concern exists. Still, we can find a point of general agreement to start our discussion. As such initial point I am going to advance a quite simple supposition: Bioethics as a field of concerns, deliberations, and systematic studies is more advanced and consolidated than is the ethics of biotechnology. Ethical issues of biotechnology became an object of special, systematic interest after bioethics was already a well-established field of research.

We witness so much disagreement among ethicists concerning virtually every aspect of bioethics that our perception of the field is not so much a field of approved and verified knowledge as a field of acute debates and contentions.

One of the premises of these discussions, which have continued for more than three decades, is that they can take place only if some commonly shared assumptions and presuppositions exist. In the course of the discussions these commonly shared assumptions became more articulated; these discussions allow us to strengthen and elaborate our shared assumptions and presuppositions. Even if people have different, sometimes contradictory beliefs and values regarding, say, the moral status of the human embryo, or ethical permissibility of human cloning, they can still engage in rational discussion of the controversies, but only because they share a common understanding of these topics.

A recent example is illustrative: To date, the Steering Committee on Bioethics of the Council of Europe (CDBI) has not succeeded in preparing a legally binding document devoted to the protection of human embryos and fetuses because they have not reconciled major disagreements. The main reason for the split is irreconcilable contradictions in understanding of moral status of human embryos. But they were able to create another agreed upon document—the report in which these exact disagreements are presented.[1] The report, was:

> [the] result of several years' work by the Council of Europe . . . highlights the numerous different approaches taken in Europe. . . .The various arguments supporting the different existing positions in Europe are clearly outlined.[2]

So, over the course of protracted discussions, the committee achieved a shared base of knowledge.

Keeping in mind that bioethics is something more elaborated and more definite than ethics of biotechnology, we can use the first as a kind of benchmark for analysis of the second. The term biotechnology—and consequently ethics of biotechnology—has many different interpretations. The *Encyclopedia of Bioethics* states that biotechnology "includes any technique that uses living organisms to make or modify products, to improve plants or animals, or to develop microorganisms for specified uses."[3] According to this broad interpretation, the ethics of biotechnology covers the entire problematic area of bioethics. So, in this sense, we have an area of overlapping between bioethics and the ethics of biotechnology.

Other interpretations draw distinctions between the two. One area of concern looks at ethical issues concerning uses of biotechnology not immediately applied to human beings, for instance, agricultural use of biotechnology, ecological problems generated by its use, and influence of biotechnologies on biodiversity. At the same time, bioethics addresses issues directly related to human beings, such as caring, which has traditionally been the subject matter of medical ethics or health care ethics. This distinction between the ethics of biotechnology and bioethics, while by no means the only existing conceptualization, appears logical.

So this understanding of the ethics of biotechnology, not so broad in its scope, is limited to only those uses of technology that do not touch directly upon human beings. This distinction appears to be a common approach in considering ethical matters related to biotechnology, and it corresponds to the general usage, which then lends the word bioethics for ethical accounts of applications of technologies directly applied to human beings. This distinction between bioethics and ethics of biotechnology is also intuitive.

We commonly distinguish between technological impact on human beings and technological impact on things in the outer world. We usually call the first type of impact intervention. Intervention can be of physical or psychical nature, and is, but are not always, directed toward achievement of medical treatment or enhancements such as the use of growth hormones or cosmetic surgery. Moral assessment of these impacts is a problem area of bioethics. Ethical accounts of interventions address the intention to care for persons or other living creatures in need or who feels pain or suffering. Our initial reaction to such situations is emotional—feelings of compassion and empathy with the persons in need, and intend to relieve pain and alleviate suffering. So, we need to assess interventions in terms of risks and benefits imposed by them to persons. The ethical analysis evaluates how well the intervention preserves and respects a person's dignity and autonomy, the accessibility of the intervention to a person, and whether equitable access to the intervention exists for all persons in need.

To distinguish impacts made by biotechnology from interventions, I call them intrusions. Intrusions are directed toward the environment around human beings. Intrusions exert influence, often profound, on human beings living in environment modified by them, but the influence is indirect or mediated.

2. Internal and External Considerations

At an intuitive level, we can easily discern interventions on us from intrusions into world around us. The distinction is analogous to the two kinds of "givenness," which we experience differently. Internally given experience, given to each individual with certainty and authenticity, can be either cognitive or emotional in nature, but in no way do these qualities preclude the possibility or necessity of subsequent rational assessment of those experiences. Our emotions, not rational arguments, generate the deeply rooted consensus or disagreement characteristic of bioethical discussions.

Externally given experience, is given first through perception of some rationally constructed data. Some kind of theoretical account, however loose, allows us to grasp that some events and processes in world around us, triggered by our biotechnological intrusions, could cause some adverse effects on us. At the same time, in the case of internal givenness, we do not need deep deliberations to grasp that, unlike an intrusion into events and processes taking place in the external world, any immediate intervention into us is fraught with possible risks, benefits, and dangers.

Any intrusion that causes an indirect effect on us might be fraught with possible risks, benefits, or dangers. But our internal perception of the intrusion may not match the reality—we may not cognitively or emotionally appreciate the danger to us—because the reality may not immediately impinge on our senses. We should not be deceived by our internally given experience because it may not match the externally given data.

We know, for instance, that after the Chernobyl nuclear power accident in April 1986, many inhabitants of radioactivity-contaminated areas did not want to leave their place of residence. One of the main reasons was that they could not properly perceive the risks and dangers related with such an indiscernible entity as radioactivity, the effects of which only became apparent after some time.

We can also compare interrelations between intervention and intrusion described in such a manner with interrelations between time as a form of inner sense and space as a form of outer sense discussed in Immanuel Kant's *Cri-*

tique of Pure Reason. For Kant, time, as inner contemplation, has priority over space. We can understand intervention, which we first experience in an internal way, as something correlative with time as a form of inner sense, which we perceive more immediately, more vitally, if you wish, than intrusion as something correlative with space, which is more remote.

Unlike the case with interventions that intentionally designed to execute well defined impacts on the human subjects, no such concentrated and distinctive intention directs our ethical concerns in relation to biotechnological intrusions. Typically, more a diffuse intention such as a concern for general systemic coherence, interdependence, and orderliness of the cosmos characterize intrusions.

The ancient Greeks perceived the cosmos to be the whole world around us, including the natural and the social world. Biotechnological intrusions mainly touch us not as individual persons, but as members, or representatives of a community that includes human and non-human members. Ultimately, in the course of interventions, our ethical concerns will push us into the depths of human inner experience, while intrusions will direct us to the breadth of the outer world.

We need to seek grounds for moral assessment of these biotechnological impacts in our knowledge about the world outside us and in our relations with this world. This knowledge and these relations get their expression in such fields of moral reasoning as ethics of natural science and technology, or environmental ethics.

To summarize, ethical concerns regarding internally given versus externally given experience have different a nature and different directedness. Consequently, we can understand bioethics and ethics of biotechnology as two distinct ways of moral deliberation. Bioethics ultimately discusses diverse ethical issues at the level of single persons, or, more generally, of single living beings, and their interests. Ethics of biotechnology is concerned with the welfare of more generic levels, such as populations or species. This differentiation does not mean that an impenetrable borderline between these two kinds of ethical analysis exists. Instead, we must understand them as ideal types in Max Weber's sense.

Viewing and assessing biotechnological developments through prism of bioethics is possible and useful. In such case we treat ethics of biotechnology as prolongation and expansion of bioethics, when, say, we treat our planet, or biosphere, or an ecosystem as a kind of a single individual. Such an approach would cause us to miss some essential ethical problems of biotechnology. Vice versa, we can productively approach (with the same reservations) many problems of bioethics from the point of view of ethics of biotechnology. In these cases, we try to perceive them in an impartial, alienated, and less emotional manner. Even if such a change of perspective would allow us to get more unbiased, balanced decisions, these decisions would lose an essential part of mobilizing potential crucial to their implementation.

Public perception of bioethical issues, as a rule, appears to be more energized by emotion than is the public perception of issues in the ethics of biotechnology. As a result, medical ethics appears more magnetizing and more urgent than the ethics of science and technology.

3. Preservation and Safety

The ethics of biotechnology address interrelations between an individual and the world, and the place and destiny of human beings in the cosmic order. We can identify two different value orientations in perceiving these interrelations and in evaluating the meaning of our intrusions. The first stresses the value of preserving the surrounding order of things, which we could easily and irreversibly destroy by rash and unreasoned actions. These motives are especially evident in tackling ecological problems arising in the course of biotechnological intrusions, such as introduction of genetically modified organisms into environment and necessity of its protection. This concern about preservation of life on the Earth that was pre-eminent in Van Rennselaer Potter's initial bioethics project.[4]

Values of preservation also dominate in a much more widely accepted concept of bioethics which goes back to Andre Hellegers, founder of the Kennedy Institute of Ethics in Georgetown University, and Daniel Callahan and Willard Gaylin, founders of the Hastings Center—in this case at stake is preservation of human life, health, rights, dignity, and autonomy. To an essential degree, this bioethics is the ethics of protecting some existing entities, especially in the face of new and often aggressive biomedical technologies. For the sake of preservation, we need to produce many changes; yet we direct all these changes toward restoration of some impaired (presumably natural) conditions, states, structures, processes, or functions.

So we see that bioethics has much in common with ecological ethics—the main concern for both of them is dangers, which can arise as results of our impacts, either interventions or intrusions, usually positively motivated, on the natural course of events. These intentions can be discerned in so called precautionary principle, which was a groundbreaking foundation for directives issued by the European Commission in 1990 about the use of genetically modified microorganisms (Directive 90/219) and about releasing into the environment genetically modified organisms (Directive 90/220). According to Francis Fukuyama, these directives do not mention this principle as such, "but their language is not inconsistent with it. The first explicit mention of the precautionary principle is made in the Maastricht Treaty of 1992."[5] The precautionary principle is used as means for the proper distribution of responsibility between proponents and opponents of new biotechnologies. This principle holds that proponents have the responsibility to prove the safety of every potentially dangerous technology.

In my opinion, the precautionary principle expresses the essence of ethics of biotechnology if ethics of biotechnology is understood as embodying value of protection, preserving world around us.

> The United States has not accepted the precautionary principle as a risk standard, arguing instead that the burden of proof must lie with those who claim that safety or environmental harm exist, instead of with those who claim they do not.[6]

Recently, the problem of safe use of new biotechnologies and precautionary principle has received new expression due to dangers of bioterrorism. We usually presume that biotechnologies are conceived and developed for the benefit of people (biological weapons present a special case beyond the scope of this discussion), however intricate the share of benefits may be. Consequently, any misuse or abuse of these technologies is the result of a lack of precaution on the parts of their creators, who were either reckless or not able to foresee possible adverse effects. These adverse effects are not deliberate—those who developed such technologies did not intend to use them for destructive ends.

But in case of bioterrorism we have exactly such inclination and use of human imagination. Accordingly, the potential for dangers related to new biotechnologies has become much wider. Consequently, we must widen the scope of our defensive imagination to counter this new source of threats. This is a brand new direction for deliberations on biosafety. We need to elaborate special means and mechanisms for the systematic assessment of new biotechnologies to make them not just foolproof, but also evil-intent-proof.

4. Changes and Hope

Now I can turn to the second value orientation in relation to new biotechnologies. This value system has recently become increasingly visible. It stresses values related with (presumably progressive, positive) changes in the existing (natural) order of things. Our interests, desires, and dreams direct these changes. This kind of ethics of biotechnology upholds our interests and desires above the imperatives of the preservation of the nature around us. We perceive nature first as a raw material at our disposal for transformation by means of our technologies to achieve our goals.

Such shift from values of preservation to values of change is discernible not only in attitude toward nature around as, but also to our own human nature. So, this opposition of values of preservation and values of change has relevance to an ethical account of applications of biotechnology to all living creatures, and to biotechnological manipulations with humans and to biotechnological interventions.

Current debates on therapy vs. enhancement of human beings reflect the opposition of these two sets of values.[7] By therapy in this case I mean a restoration of human health which can be interpreted as a preservation, instead of modification, of existing human nature, whereas enhancement implies change, usually improvement or advancement. We can find expression of this opposition in the realm of ethics of biomedical research. In its more traditional forms, ethics of research stressed risks and burdens for the participants. In each case, the involvement of human beings in biomedical research is a risky endeavor that needs to be scientifically justified and ethically approved. A researcher has an obligation to guarantee minimal or acceptable level of risk for participants. The participants, in turn, have a right to choose whether to participate or not. The person in question decides whether to use ordinary, existing, approved methods of therapy, consequently, to preserve the current state of the art, or to promote a search for new methods that constitute change.

More recent versions of research ethics stress benefits for persons, which they stand to receive because of their participation as subjects, because they will have access to any progress in therapy identified by the research. Some authors hold that individuals have an obligation to be personally involved in the promotion of change. John Harris holds that "where risks, dangers or inconvenience of research is minimal, and the research well founded and likely to be for the benefit of oneself or others, then there is some, perhaps quite modest, moral obligation to participate," and "To fail to contribute to research is against the public interest and may harm others."[8] Harris's argument is built on the premise that our participation in research contributes to the overall welfare of the community. He presupposes that the community can achieve benefits by the way of some changes, namely, manipulations with a person, or with that person's personal data.

Many may view the biotechnological creation of new organisms as a presumptuous disregard for the world created not by us, the world in which we are only temporary inhabitants. People with different ethical views and preferences will treat this problem differently. Our existence in this world would be impossible without constant and essential interruptions of "naturally" occurring sequences of events. Yet our demiurgic potencies and abilities must not be overestimated—after all, the world around us deserves our respect.

NOTES

1. Steering Committee on Bioethics (CDBI). The Protection of Human Embryo *in vitro*. Report by the Working Party on the Protection of Human Embryo and Fetus (CDBI-CO-GT3). Strasbourg, 19 June 2003.

2. Abstract of the 24th Meeting of the Steering Committee on Bioethics of the Council of Europe, Strasbourg, 17–20 June 2003.

3. Nanette Newell, "Biotechnology," in Warren Reich, ed., *The Encyclopedia of Bioethics*, rev. ed. (Washington, D.C.: Simon & Schuster, 1995), p. 283.

4. Ibid., p. 199.

5. Ibid.

6. Ibid.

7. See Leon R. Kass, "Ageless Bodies, Happy Souls: Biotechnology and the Pursuit of Perfection," *The New Atlantis* (Spring 2003), pp. 9–28; United States President's Council on Bioethics, "Distinguishing Therapy and Enhancement," Staff Working Paper (9 June 2004) http://www.bioethics.gov/background/workpaper7.html (accessed 2 August 2005).

8. John Harris, "Research Ethics Committees—the Future," *Politeia*, 18:67 (2002), p. 128.

Part Three

VULNERABILITY, POWER, AND RESPONSIBILITY

Eleven

INFECTIOUS DISEASE AND THE ETHICS OF RESEARCH: THE MORAL SIGNIFICANCE OF COMMUNICABILITY

Leslie P. Francis, Margaret P. Battin,
Jeffrey R. Botkin, Jay A. Jacobson, and
Charles B. Smith

1. Research and Communicability: The Problem of Risks to Indirect Participants

Do infectiousness and communicability raise unrecognized problems for informed consent in the ethics of research? Even when research is otherwise justifiable on scientific grounds, we argue here, where research involves the possibility of communication of disease, we should not understand informed consent solely as a matter of the consent of the individual subject. Others might be infected, put at risk of infection, or left unprotected from these risks. These third parties have interests that research may directly affect. Yet, as we will show, current policies about informed consent, criticisms of problematic examples in the history of medical research, and current research practice largely ignore these third-party issues. If research policies or practices attend to such issues at all, they do so principally through the direct subject—for example, through a consent process that warns the direct subject about the risks of communicability.

Taking issue with past and current practices, we argue that this focus on the direct subject is inadequate in a significant range of cases of research involving communicable infectious disease. We defend the following claims:

(1) Third parties who are at foreseeable and direct risk of contagion when research with human subjects involves a communicable agent are, in some respects, analogous to direct subjects of the research. Although they are not *subjects* in the sense that data is being collected about them, they are indirect *participants* in the research in that the research puts them at risk—or fails to shield them from risk—in the same way direct subjects are put at risk. We have chosen the term *indirect participants* to describe these affected third parties, in contrast to *indirect subjects*, a term that implies they are fully subjects, or *indirect objects*, a term that

ignores agency. The role of indirect participants as autonomous agents with respect to the research is ethically relevant, and failure to consider it is ethically problematic. Solutions to this problem are complex.

(2) When the research involves risks that are comparable to the experiences people would otherwise have, and that are only minimal risk, the approach should be to inform the research subject about the risks and to recommend that the research subject inform others about the risk. The investigators need not inform third parties directly if the risk posed by the research to them is minimal. An increased risk of contracting a disease with significant morbidity or mortality would not fall into this category.

(3) When the risk involves more than minimal risk and is not commensurate with the experiences of ordinary life, what further contact with indirect participants is required is contingent on whether the indirect participant can readily avoid the risk. If the risk is readily avoidable, for example by abstaining from sexual contact, then informing indirect participants of the risk is sufficient. The indirect participant can then choose whether to take the risk. Such action to inform the indirect participant requires the consent of the direct subject. If the direct subject refuses to permit researchers to inform the indirect participant, the direct subject should not participate in the research. If the risk is not readily avoidable by the indirect participant, researchers should be required to obtain informed consent from indirect participants before direct subjects can ethically participate in the research. An example of this category would be testing a new vaccine for a serious disease, where a significant possibility of viral shedding that might infect other members of the subject's household exists.

These claims raise a host of questions about informed consent more generally. For example, what constitutes ethical informed consent if a household member at risk is a child, an adult with disabilities, or a patient with dementia? Will consent of the direct subject be sufficient? Which individuals compose the population of indirect participants? What constitutes ethical practice if research poses risks to entire communities? What are the parallels between research with infectious agents and other research involving subjects who may be dangerous to others—violent subjects, for example—but not because they are contagious?

2. Current Informed Consent Policies and Contagiousness

Guidelines for research involving human subjects uniformly fail to address adequately whether research involving the possibility of communicable conditions requires attention to indirect participant information or consent. Current guidelines typically first analyze whether the research is permissible based on its scientific merit and risk/benefit ratio, and then establish standards for inclusion of subjects, minimization of risks to subjects, and informed consent. The analyses of scientific merit and risk/benefit ratio address the research overall instead of the situations of individual subjects. The inclusion standards and informed consent protocol consider risks to individual direct subjects. The requirement of informed consent protects individual direct subjects from involuntary participation in research. The World Health Organization Declaration of Helsinki, the United States *Federal Regulations Governing Research with Human Subjects*, and the *Guidelines of the Infectious Disease Society of America* all fit this pattern.

The Declaration of Helsinki specifies that research must meet criteria of scientific and ethical adequacy. The most general requirement specifies caution about risks in research design: "Appropriate caution must be exercised in the conduct of research which may affect the environment, and the welfare of animals used for research must be respected."[1] With respect to inclusion of any human subjects in the research at all, the declaration requires "careful assessment of predictable risks and burdens in comparison with foreseeable benefits to the subject or to others."[2]

Addressing the inclusion of individual subjects, the declaration shifts focus, prohibiting their involvement in research unless the "importance of the objective outweighs the inherent risks and burdens to the subject,"[3] without mention of risks or benefits to others. Informed consent is required from "each potential subject." Required information researchers must supply to subjects includes the aims, methods, sources of funding, any possible conflicts of interest, institutional affiliations of the researcher, the anticipated benefits and potential risks of the study, and the discomfort it may entail.[4] This list does not include explicit mention of risks to others. Instead, the declaration directs attention only to population risk/benefit analysis in determining the permissibility of the research overall, and to risks to direct subjects in determining the permissibility of their individual inclusion in the research.

The *Federal Regulations Governing Research with Human Subjects* describe risks and benefits researchers should consider in determining whether research is acceptable. The regulations require "risks to subjects [be] minimized" by "sound research design" and procedures "which do not unnecessarily expose subjects to risk."[5] Research risks must be "reasonable in relation to anticipated benefits, if any, to subjects, and the importance of the knowledge that may reasonably be expected to result."[6]

Under the regulations, Institutional Review Boards (IRBs) must review research with human subjects. This review should consider only risks and benefits that may result from the research and not risks and benefits of therapies subjects would receive outside of the research. Possible long-range effects of applying knowledge gained in the research (for example, the possible effects of the research on public policy) are not within IRB responsibility.[7] The regulations require the informed consent process to include "a description of any reasonably foreseeable risks or discomforts to the subject" and "a description of any benefits to the subject or to others which may reasonably be expected from the research."[8] Notably, this list explicitly mentions risks and benefits to direct subjects. With regard to others, including indirect participants, the list mentions benefits but not risks.

In 1993, the Infectious Disease Society of America published guidelines for ethical conduct. The guidelines for research specified that all research involving human subjects must be subject to IRB review. They charged IRBs to "provide a framework for sound scientific work, while at the same time acting as advocates of the rights of patients, human subjects, and experimental animals and of the public welfare."[9] The guidelines did not explain whom we might regard as experimental subjects or address whether researchers should consider informed consent from indirect participants. The guidelines' silence on these matters might stem from their original purpose, responding to identified abuses of direct subjects in research studies such as the Tuskegee syphilis study.

3. Historical Examples of Ignoring Contagion: Tuskegee and Willowbrook

Some of the more (in)famous examples of research with human subjects have involved infectious diseases, although ironically, the aspect of infectiousness was not at the forefront of the criticism. We hypothesize that coincidence alone does not explain why research projects involving infectious disease have been among the more notorious examples of ethically problematic research. We postulate that some researchers have held the importance of protecting the public health to provide an overriding justification that eclipsed the ethical importance of informed consent in research. Others have sharply criticized this research history as abusive of direct study subjects.[10] Criticism has virtually ignored the issue of informed consent for indirect participants. The overwhelming nature of the concerns about the treatment of direct study subjects may have swamped concerns about indirect participants. But in the approach to informed consent that we criticize below, researchers and critics have focused on the autonomy of direct subjects while ignoring potential risks to others who may be directly affected by the research.

The Tuskegee syphilis study largely fits this paradigm of criticism. The most widely read history of the study, James Jones's *Bad Blood*, describes Nurse Rivers, the nurse who played a long-term role in the study, as going to the homes of study participants and providing general health care to their families. Yet Jones's description contains no mention of the ethical issues raised by the possibility of transmission of syphilis or of the possible impact on family members if they contracted the disease.[11]

The principal compendium of documents on Tuskegee, Susan Reverby's *Tuskegee's Truths*, reflects minimal reference to effects on indirect participants such as sexual partners—even though, at the time the study began, one justification for treatment of latent syphilis was the risk that patients might infect others. The Public Health Service, in deciding to observe the natural history of the disease in those with latent infection, assumed it was dealing with subjects who would otherwise not receive treatment. Initial testing indicated lower rates of infection in Macon County than had been anticipated, at 20 percent instead of 35 percent. Researchers told participants they were receiving "treatment," so understandably they might not have anticipated their risks to others. They were not informed of these risks of contagion, and they did not receive information about risks to their sexual partners—partners who, given attitudes about race and sexuality at the time, surely were anticipated to be at risk.[12]

The study design did not include tracking or reporting whether subjects infected others.[13] Researchers did not consider transmission of congenital syphilis to offspring. Jean Heller, the reporter who broke the story about the study, labeled the subjects as the "victims" of the study.[14] The Ad Hoc Study Panel appointed by the United States Department of Health, Education, and Welfare (HEW) to investigate the study concluded, in 1973, that researchers had wrongfully failed to inform subjects about their risks to others. They drew no further conclusions about wrongs to third parties.[15] The program of compensation for study victims did include women and children infected by study participants. Neither United States' Congressional inquiries nor United States' President William Jefferson "Bill" Clinton's apology on behalf of the nation referred to potential third-party victims.[16] Amy L. Fairchild and Ronald Bayer's recent critique similarly focuses on the direct subject:

> [T]hree critical features that characterize the nature of the consistent research abuses that occurred over the course of forty years. The study involved, first, deceptions regarding the very existence and nature of the inquiry into which individuals were lured. As such, it deprived those seeking care of the right to choose whether or not to serve as research subjects. Second, it entailed an exploitation of social vulnerability to recruit and retain research subjects. Finally, Tuskegee researchers made a willful effort to deprive subjects of access to appropriate and available medical care, which changed over time, as a way of furthering the study's goals.[17]

The third-party concerns focused on the general public health effects of the study, not on the possibility of harms done to indirect participants. Allan Brandt wrote, for example, "The entire health of a community was jeopardized by leaving a communicable disease untreated."[18] This public health concern, while laudable, is a different concern from the one we raise here. We are concerned that researchers violated the autonomy of indirect participants because they did not afford them—in the study design or in subsequent apologies—the choice whether to undertake exposure to the study's risks.

The much-criticized Willowbrook research is a similar example. That project used residents of a treatment facility for individuals with developmental disabilities to study the transmission of infectious hepatitis. The investigators undertook the research because the infection rate among residents at the facility was high. They reasoned that infectious hepatitis was "mild and relatively benign" in children by comparison with adults.[19] Reflecting back on the study, its author concluded that the infection rates among patients and employees in comparison to what were judged minimal additional risks to subjects justified the study.[20] The study design isolated children who received the artificial hepatitis infection from exposure to other infectious diseases common in the institution.[21] Informed consent was required from the parents of the children involved on behalf of the children. Subsequent extensive criticism of the Willowbrook study design and consent process does not appear to have addressed risks of transmission to the parents or siblings of the children or staff of the institution, or the need to involve these parties on behalf of themselves in the informed consent process.[22]

4. Historical Examples of Considering Contagion: The Common Cold and Polio Vaccine

Several other historical examples of research did place communicability at the center of concern. Research involving transmission of the common cold and research on the polio vaccine are noteworthy examples.

Early research involving transmission of the common cold considered the risks of transmission in the study design. Studies performed at the Common Cold Research Unit in Britain during World War II attempted to ascertain whether colds are transmitted via nasal secretions and whether exposure to freezing temperatures is a risk factor. Conducted at an isolated research facility, these research designs minimized the risk of external transmission while guarding against contamination of treatment groups. Later studies, addressing the efficiency of manual transmission and issues such as the effect of colds on memory, did not feature the two-week isolation characteristic of the British experiments.[23]

Researchers developing polio vaccines considered the risk of contracting polio to the person receiving the vaccination. With the killed-virus vaccine (IPV), researchers were concerned whether the vaccine consisted of imperfectly killed virus or would prove ineffective. The risk of imperfectly manufactured research material soon became apparent when vaccine produced by

Cutter Laboratories proved to contain live virus that resulted in a number of cases of polio.[24]

The live-virus vaccine (OPV), which causes viral shedding from the vaccinated person, has the advantage of increasing population immunity but the disadvantage of risking infection to the vaccinated patient and to others exposed to the vaccinated patient. Discussions of the ethics of using OPV indicate that everyone involved was well aware of these risks. Jonas Salk persistently championed IPV because of its lower third-party risk profile. In 1979, arguing for the use of IPV, he pointed out that in the United States during 1969–1977, records reported twenty-four cases of OPV-associated paralytic polio among vaccine recipients, forty-seven cases among direct contacts of vaccinated people, and sixteen among indirect contacts.[25] Advocates of OPV argued that OPV would produce higher levels of population immunity resulting in fewer cases of paralytic polio overall than IPV.[26] A 1977 report by the Institute of Medicine Committee for the Study of Poliomyelitis Vaccines concluded that OPV was the preferred method of immunization but that IPV should continue to be offered to people with heightened susceptibility to infection and to people who prefer IPV and are prepared to make a commitment to the required full schedule of vaccinations. About informed consent, the report concluded that documents should be "as brief as possible, while conveying sufficient, accurate information on the benefits, risks, and other special characteristics of the vaccines." The report also recommended that the United States' federal government should assume the responsibility to compensate people with vaccine-associated poliomyelitis, including contact cases.[27]

Notably, though, these discussions did not attend to the problems of informed consent for indirect participants at risk of contagion. Because many of the subjects were children, whose parents consented on their behalf may complicate this assessment since the parents would become aware of the risks as they consented to inclusion of their children in the research.[28] Still, there were no indications that researchers asked parents to consent on behalf of their other children who might be subject to exposure but not receiving vaccine. The assumption that the background risk of contracting polio was quite high in any event, especially for children, might also have complicated this assessment.

The first example of direct attention to the problems of research ethics raised by indirect subjects in polio vaccination was Raisa B. Deber and Vivek Goel's 1990 analysis. They pointed to the ethical difference between polio cases that occur to persons who lack immunity, polio cases that involve people who have consented to vaccination, and polio cases resulting from transmission by direct or indirect contact to those who have been vaccinated. Cases in the last category, they contended, involve a kind of involuntary immunization without consent for persons who are susceptible precisely because they have chosen not to be immunized in the first place.[29]

These issues are ongoing in contemporary public health debates about polio eradication.[30] Similar controversy may be emerging over the development of a vaccine against avian flu.[31] Concern about risks to immediate contacts also played a role in controversy over recent efforts to immunize first responders against smallpox. The Institute of Medicine has advised caution in extending the immunization program until risks are further evaluated, including risks to contacts.[32]

5. Contemporary Examples of Ignoring Contagion

Many contemporary examples of research involve contagious conditions. A principal source of contagion for indirect participants occurs when the experiment is testing a method to prevent infection that proves ineffective. In other cases, the research features an intervention that may increase risks of contagion, or the research introduces a new possibility of infection and subsequent contagion. Yet examination of selected examples of current practice evinces little attention to information or consent from anticipated third parties.

Trials of a vaccine against herpes simplex and trials of short-course antiretroviral therapy provide examples of contemporary studies where researchers have not included indirect participants in the consent process. Investigators have addressed third party risks specifically in studies of xenotransplantation, the surgical implantation of cells, tissues, or organs from one species to another. Another example of high profile research where critics have raised questions about risks of infection and transmission, but which we do not consider here, is gene therapy, where the mechanism of delivery is a viral vector. In those cases, concern appears focused on direct subjects, but not on their direct contacts who, we might regard as indirect participants.[33]

GlaxoSmithKline, a global pharmaceutical corporation, has sponsored several clinical trials of the safety and efficacy of a vaccine against the herpes simplex virus. The trials enrolled over 400 adult women at risk of contracting herpes, in over eighty study centers on four continents (Africa, Australia, Europe, and North America).[34] The consent form detailed an extended list of risks, including risks of the injection and of blood draws that would be required to test levels of immunity. The form explains in detail that the vaccine contains bovine derivatives manufactured to avoid any risk of bovine spongiform encephalopathy ("Mad Cow" disease). Bold face type cautions participants to avoid pregnancy, as the effects of the vaccine on a fetus are unknown. In an open label study, which offered active vaccine to patients who had received placebo in an earlier trial, researchers informed patients that 73 percent of women receiving vaccine in the earlier study had been effectively immunized against herpes simplex. Disclosures reminded participants that the vaccine was still experimental and that they might receive no benefit from study participation, but the consent form contained no mention of sexual partners or of the risk that if the vaccine proved ineffective and the subjects acquired herpes infection, they might then pass the infection on to others.

Regardless of actual participation, potential participants in this study were at background risk of contracting herpes and transmitting it to their partners. Participation offered a possibility of protection because of the vaccination that they would not otherwise have had. Their sexual partners, defenders of the study could claim, were not at any greater risk from the study than they otherwise would have been, and may even have been at reduced risk, so any discussion of risks to them is arguably unnecessary.

This argument has two related replies. First, research studies are held to higher standards than the background risks that might have occurred regardless of research participation. Risk/benefit ratios must be acceptable and researchers must minimize risks to subjects if possible. We contend that this minimization of risks should apply not only to direct subjects but also to others who might be immediately affected by the subjects' participation, which makes them indirect participants in the research. Second, participants in the study may have believed that they were receiving protection, when in fact they were not. Believing they had received adequate protection, they might have been less careful to guard themselves and their partners against risks of contagion. These concerns, we contend, at minimum require including in the consent form the reminder that the vaccine might not be effective, and that *if* it is not, participants *and* their partners may be at risk of infection.

Trials of short-course anti-retroviral therapy to reduce the likelihood of transmission of HIV from pregnant women to their fetuses have been highly controversial. Participants in these trials have been HIV-positive women in areas of the world where access to anti-retroviral therapy is limited or nonexistent. Because short-course therapy appeared economically feasible, whereas optimal therapy did not, trials have been designed to determine whether short course therapy significantly reduces vertical transmission rates.

Fifteen different trials have compared vertical transmission rates in patients receiving the short-course therapy with vertical transmission rates in patients receiving placebo.[35] Defenders of the trials argued that the researchers' duty to care for subjects did not extend beyond the best standard of care available in the subjects' circumstances. Critics argued that because there was an alternative to placebo—the standard of care available in the developed world—the studies violated clinical equipoise (the requirement that studies compare therapies not known to be superior or inferior to each other) and were clearly morally wrong, comparable to Tuskegee in observing seriously ill patients for whom treatment was possible.[36]

Bioethicists have raised many serious ethical issues about these trials. Alex John London argued that the concept of equipoise is ambiguous between a narrow physiological concept and a broader concept of efficacy in social context.[37]

Solomon R. Benetar has argued for sweeping reforms in the understanding of ethics in international research, in the context of international justice in health care.[38] Many discussions have highlighted coercion and exploitation of vulnerable populations.[39] In addressing informed consent from study subjects, the principal suggestion has been involvement of families or communities in the consent process. This suggestion has rested on several grounds: Family and community support is an important protection against exploitation. The societies in which the studies have taken place frequently are communitarian in structure, and do not see consent as an individualized process.[40] The studies may have effects on the communities in which they take place.

More recently, concerns have surfaced about the risks of short-course anti-retroviral therapy to the women involved. One risk is that if anti-retroviral therapy becomes available, later treatment efforts may prove less efficacious for women who have previously received the short-course therapy. Another risk is that the short-course therapy may alter the course of the subjects' disease, generating increased viral load or more resistant viral strains. Either of these risks are direct risks to the study subjects. But given levels of heterosexual transmission of HIV, they are also risks to the subjects' sexual partners.[41] Yet we have found *no* criticisms of the HIV trials that mention the unacknowledged increased risks to partners as an ethical issue in the studies. Where we did find mention of family members, the focus was not on risks to them, but embedded in familial or communal informed consent models.[42]

Unlike the studies that are silent regarding third-party issues, protocols studying xenotransplantation have undergone extensive scrutiny for the safety of individual participants and the public. Perhaps the difference stems from perceptions that this procedure is new and strange and fears of species-jumping infections of unknown character. Margaret A. Clark, for example, argues that patients receiving xenotransplants should be required to consent to take precautions against transmission of bodily fluids. She also argues—a view we share—that transplant patients and their intimates must receive information about transmission risks. She suggests that researchers should seek population consent where there are significant public health risks, but she does not otherwise raise the possibility of indirect participant consent where direct risks to third parties are apparent.[43]

6. Considering the Risks to Indirect Participants

In the historical and contemporary examples we have discussed, participation of direct subjects in the research creates potential risks for their immediate contacts, such as family members or sexual partners, and sometimes for the public at large. Public policy requirements, study design and approval, and criticism of studies have addressed risks to individual subjects and to society generally. Research practices are considered unethical if they include direct subjects into studies without informing them of the risks they face and consequently without providing information relevant to decisions about participation. But close contacts of such subjects—sexual partners or family members in particular—might experience exposure to risks by the participation of direct subjects as well. Yet research designs, policy statements, and critics alike virtually ignore these risks to indirect participants. The failure to consider such third party risks extends even to inattention to providing information about risks of infectiousness to study subjects. This gap in policy and practice, we contend, represents a failure to respect the autonomy of third parties that is analogous to the failure to respect the autonomy of the direct subjects. In response to these concerns, we suggest several changes in current policy:

(1) Attention to risks to indirect participants in study design: Study designs should explicitly consider risks of third-party transmission. Ensuring that the study has a risk/benefit ratio that is favorable overall will not accomplish this goal. Ethical study design requires not only minimization of risks to subjects, but also minimization of risks to indirect participants. The likelihood of creating indirect participants might be reduced by strategies such as isolating the direct subject until the likelihood of contagion has passed. Otherwise, people may be subject to risks of the study that might have been avoided in the study design, even though they are not themselves subjects of the research. In some studies involving contagious conditions, design features such as these may not be feasible. For example, research with OPV created immunized subjects who continued to shed potentially infectious virus for extended time periods.[44]

(2) Informed consent with direct subjects: When applicable, the informed consent process should include information about risks of contagion. Issues addressed with direct subjects should include risks of contagion that might result from their participation in the study. Researchers should warn subjects in prevention studies, for example, that if the method of prevention fails and they become infected, they might pose a risk of contagion to others. In studies that may raise subjects' risks of contagion, such as the trials of short-course anti-retroviral therapy, researchers must inform subjects of this possibility so that they may consider it when deciding whether to participate in the research.

(3) Informed consent with indirect participants: Current practice relies on direct subjects to provide contacts with information as in the herpes or xenotransplantation research. If the risk to indirect participants is substantial and serious, this practice may be insufficient. Direct subjects may be reluctant to convey the information. They may understand the risk imperfectly or may be unable to explain what they understand. In such circumstances, con-

tacts—the indirect participants—may not receive the information they need to protect themselves should the infection risk eventuate. Protecting the autonomy of the indirect participant, we contend, requires more.

At a minimum, any information concerning significant and serious risk should directly reach indirect participants, so that they can understand and act on it. If the risk is unavoidable, we should also consider whether respect for indirect participants requires their independent consent to the direct subject's participation in the study. In short, when risks to indirect participants are potentially as great as risks posed to direct subjects of research, we recommend that researchers afford respect for the autonomy of the indirect participant similar to that afforded to direct subjects. They should allow contacts or family members of a person participating in infectious disease research to understand that the direct subject may be victim and vector of the disease under study.

We recognize that our approach limits choices open to the direct subject. Direct subjects who do not wish indirect participants to be informed about the risks of the study will not be able to participate in the research. If the indirect participant refuses consent, when consent is required, the direct subject's participation will also be blocked.

We think that this approach strikes a defensible balance, though. Despite the standard that encourages subjects to base their decisions about participation on a reasonable assessment of the risks and benefits to them, we should not equate their participation in a research study with receiving beneficial therapy. When subjects participate in research in a way that directly affects the risks and benefits for others in significant ways, but these others are not consulted, the others are affected by the research without their consent. When the risks are serious and unavoidable, as they might be with transmission of zoonotic diseases from xenotransplants, their informed consent should also be required.

7. Considering Indirect Participants: How Far to Go?

Despite the case we have made for considering risks to indirect participants, we also recognize serious issues about the potential range of indirect participants. Spreading the range too far—say, to fourth or fifth parties, the contacts of the contacts of the indirect participants—could strangle infectious disease research. We readily admit that we do not have final answers. Our goal in this chapter has been to direct attention to a significant ethical issue heretofore virtually ignored. Despite all this, we would suggest that respect for involuntarily involved third parties such as sexual partners or family members requires their involvement in the consent process when they are identifiable and are at known, direct, and significant risk.

Another difficulty with our proposals is that some indirect participants may not be able to understand information or to give their informed consent. Children and cognitively impaired family members may be at greater risk of disease transmission and unable to give informed consent. The standard way—with which we agree—to handle these situations, is to obtain informed consent on their behalf from these third parties' proxies. But we advise special caution when the proxy, for example a parent, also serves as the proposed direct subject of the research. In that case, the proxy/direct subject may not be an independent representative of the best interests of the indirect participant.

At some point, we might anticipate that concern about contagion to indirect participants might extend to concern about the community. Suppose we know that risks of rapid transmission of a serious disease such as Severe Acute Respiratory Syndrome (SARS) or avian flu exist. Research on the development of vaccines for these diseases might have an overall favorable risk/benefit ratio, given the apparent seriousness of the diseases. But in the community in which the vaccine trial takes place, members of the community might be placed at special risk, depending on study design.

For studies not involving infectiousness, where we know that the study may affect some community members, but these community members cannot be identified in advance, the model has been community consent. An example is the trial of installing cardiac defibrillators in community settings such as shopping centers.[45] We would suggest further exploration of this possibility in the infectious disease context, if research poses such special risks to communities.

The ethical problem of indirect participation in research is not an entirely novel concept. In the context of genetic information, where participation by some family members can result in the collection of information about identifiable other family members, third-party consent has been raised as a possibility when the risks to the third parties are significant.[46]

In the case of genetic information, the third parties are subjects in the sense that the research involves the collection of information about them, and researchers have regarded them as secondary subjects of the research. When studies involve diseases that are infectious and contagious, identifiable contacts may not be subjects in the sense that information about them is being collected. Nonetheless, the risks of their indirect participation may be as substantial and serious to them as the risks of collecting their genetic information.

ACKNOWLEDGMENT

Our thanks are due to our research assistant, Linda Carr-Lee, who has been invaluable in the preparation of this chapter.

NOTES

1. World Medical Association, "Declaration of Helsinki: Recommendations Guiding Physicians in Biomedical Research Involving Human Subjects" *Informed Consent in Medical Research*, eds. Len Doyal and Jeffrey S. Tobias (London: BMJ Books, 2001), pp. 4–6; rev. ed., cf. World Medical Association, "Declaration of Helsinki: Ethical Principles for Medical Research Involving Human Subjects, Edinburgh, Scotland, October 2000," *World Medical Association Recommendations Guiding Physicians in Biomedical Research Involving Human Subjects* (Ferney-Voltaire Cedex, France: World Medical Association, 2002), § B.12.
2. Ibid, § B.16.
3. Ibid, § B.18.
4. Ibid, § B.22.
5. United States Code of Federal Regulations 45 (2004) § 46.111(a)(1)(i).
6. Ibid., § 46.111(a)(2)(i) (2004).
7. Ibid.
8. Ibid., § 46.111(a)(2)(3) (2004).
9. Infectious Diseases Society of America, "Guidelines for Ethical Conduct by Members and Fellows," *Journal of Infectious Diseases*, 167:1 (1993), pp. 257–258.
10. Julie Rosenbaum and Ken Sepkowitz, "Infectious Disease Experimentation Involving Human Volunteers," *Clinical Infectious Diseases*, 34 (1 April 2002), pp. 963–971; and J. David Smith and Alison L. Mitchell, "Sacrifices for the Miracle: The Polio Vaccine Research and Children with Mental Retardation," *Mental Retardation*, 39:5 (2001), pp. 405–409.
11. James Jones, *Bad Blood* (New York: The Free Press, 1993).
12. Susan Reverby, ed., *Tuskegee's Truths: Rethinking the Tuskegee Syphilis Study* (Chapel Hill: University of North Carolina Press, 2000), esp. p. 4; and Allen M. Brandt, "Racism and Research: The Case of the Tuskegee Syphilis Experiment," Reverby, *Tuskegee's Truths*, pp. 4, 15–33, esp. 20, 21, 23.
13. Thomas Benedek, "The 'Tuskegee Study' of Syphilis: Analysis of Moral versus Methodologic Aspects," Reverby, *Tuskegee's Truths*, pp. 213–235, at p.230.
14. Jean Heller, "Syphilis Victims in U.S. Study Went Untreated for 40 Years," Reverby, *Tuskegee's Truths*, pp. 116–118.
15. Tuskegee Syphilis Study ad hoc Advisory Panel to the Assistant Secretary for Health and Scientific Affairs, "Selections from the Final Report," Reverby, *Tuskegee's Truths*, pp. 157–181.
16. William Jefferson Clinton, "Remarks by the President in Apology for Study Done in Tuskegee," Reverby, *Tuskegee's Truths*, pp. 574–577.
17. Amy L. Fairchild and Ronald Bayer, "Uses and Abuses of Tuskegee," Reverby, *Tuskegee's Truths*, pp. 589–604, at p. 590.
18. Brandt, "Racism and Research," p. 28.
19. Saul Krugman, Joan P. Giles, and Jack Hammond, "Infectious Hepatitis: Evidence for Two Distinctive Clinical, Epidemiological, and Immunological Types of Infection," *Journal of the American Medical Association*, 200 (1967), pp. 365–373. Republished (1984) as a "Landmark Article," *Journal of the American Medical Association*, 252:3 (1984), pp. 393–401.
20. Saul Krugman, "The Willowbrook Hepatitis Studies Revisited: Ethical Aspects," *Reviews of Infectious Diseases*, 8:1 (1986) pp.157–162, at p.157.

21. Ibid., p. 159.
22. Stephen Goldby, "Letter: Experiments at the Willowbrook State School," *Lancet*, (10 April 1971), p. 749.
23. Rosenbaum and Sepkowitz, "Infectious Disease Experimentation Involving Human Volunteers," pp. 965, 967.
24. Allan M. Brandt, "Polio, Politics, Publicity, and Duplicity: Ethical Aspects in the Development of the Salk Vaccine," *Connecticut Medicine*, 43:9 (1979), pp. 581–590, at p. 587.
25. Jonas Salk, "Immunization against Poliomyelitis: Risk/Benefit/Cost in a Changing Context," *Developments in Biological Standardization*, 43 (1979), pp. 151–157.
26. Alan R. Hinman, Jeffrey P. Koplan, Walter A. Orenstein, and Edward W. Brink, "Decision Analysis and Polio Immunization Policy," *American Journal of Public Health*, 78:3 (1988), pp. 301–303.
27. Elena O. Nightingale, "Recommendations for a National Policy on Poliomyelitis Vaccination," *New England Journal of Medicine*, 297:5 (1977), pp. 249–253, at p. 253.
28. Brandt, "Polio, Politics, Publicity, and Duplicity: Ethical Aspects in the Development of the Salk Vaccine."
29. Raisa B. Deber and Vivek Goel, "Using Explicit Decision Rules to Manage Issues of Justice, Risk, and Ethics in Decision Analysis: When is it not Rational to Maximize Expected Utility," *Medical Decision Making*, 10:3 (1990), pp. 181–194, at p.192.
30. Leslie Roberts, "Two Steps Forward, One Step Back in Polio Fight," *Science*, 304 (May 2004), p. 1096; Leslie Roberts, "Health Workers Scramble to Contain African Epidemic," *Science*, 305 (July 2004), pp. 24–25; Leslie Roberts, "Fighting Polio Block by Block, Shack by Shack," *Science*, 303 (March 2004), pp. 1965–1966; and Leslie Roberts, "The Exit Strategy," *Science*, 303 (March 2004), pp.1969–1971.
31. Keith Bradsher and Lawrence K Altman, "A War and a Mystery: Confronting Avian Flu," *New York Times* (12 October 2004).
32. Elizabeth Olson, "Panel Urges Shift of Focus in Preparing for Smallpox," *New York Times* (12 August 2003); and Board on Health Promotion and Disease Prevention, Institute of Medicine, "Letter Report #3" (27 May 2003).
33. Catherine S. Manno, Amy J. Chew, Sylvia Hutchison, et al., "AAV-Mediated Factor IX Gene Transfer to Skeletal Muscle in Patients with Severe Hemophilia B," *Blood,* 101:8 (2003), pp. 2963–2972.
34. Consent form, Protocol # 208141/037 (HSV-037) (December, 2001), on file with the University of Utah Institutional Review Board.
35. Paquita De Zulueta, "Randomised Placebo-Controlled Trials and HIV-Infected Pregnant Women in Developing Countries: Ethical Imperialism or Unethical Exploitation?" *Bioethics*, 1:4 (2001), pp. 289–311, at p. 292.
36. Marcia Angell, "The Ethics of Clinical Research in the Third World," *New England Journal of Medicine*, 337:12 (2001), pp. 847–849.
37. Alex John London, "Equipoise and International Human-Subjects Research." *Bioethics*, 15:4 (2001), pp. 331–332.
38. Solomon R. Benatar, "Commentary: Justice and Medical Research: A Global Perspective," *Bioethics*, 15:4 (2001), pp. 333–340.
39. Keymanthri Moodley, "Vaccine Trial Participation in South Africa—An Ethical Assessment," *Journal of Medicine and Philosophy*, 27:2 (2002), pp. 197–215; and Paquita De Zulueta, "Randomised Placebo-Controlled Trials and HIV-Infected

Pregnant Women in Developing Countries: Ethical Imperialism or Unethical Exploitation?" *Bioethics*, 1:4 (2001), pp. 289–311.

40. Moodley, "Vaccine Trial Participation in South Africa."

41. Thomas C. Quinn et al., "Viral Load and Heterosexual Transmission of Human Immunodeficiency Virus Type 1," *New England Journal of Medicine*, 342:13 (2000), pp. 921–929.

42. De Zulueta, "Randomised Placebo-Controlled Trials and HIV-Infected Pregnant Women in Developing Countries," p. 303.

43. Margaret A. Clark, "This Little Piggy Went to Market: The Xenotranplantation and Xenozoonose Debate," *Journal of Law, Medicine, and Ethics*, 27:1 (1999), pp. 137–152, at p. 141.

44. Alan W. Dove and Vincent R. Racaniello, "The Polio Eradication Effort: Should Vaccine Eradication Be Next?" *Science*, 277: 5372 (1997), pp. 779–780.

45. On file with the University Utah Institutional Review Board.

46. Jeffrey R. Botkin, "Protecting the Privacy of Family Members in Survey and Pedigree Research," *Journal of the American Medical Association*, 285:2 (2001), pp. 207–211.

Twelve

IS THERE A DUTY TO SERVE AS RESEARCH SUBJECTS?

Leonardo D. de Castro

1. Two Forms of the Question

Is there a moral duty to serve as research subjects? In light of the many ways in which we can construe this question, in this essay we will focus on the following forms:

(1) Do all persons have a duty to serve as a research subject?
(2) Are there situations when we are morally justified to ascribe a duty to serve as research subjects to particular individuals?

We will explore the extent to which we may validly claim that human beings are under some kind of moral imperative to contribute to the biomedical enterprise by serving as research subjects. We will examine how different conceptions of political, social, or contractual obligations can arguably demonstrate a moral duty to serve as subjects of biomedical research. We will discuss problems that arise when we model a duty to serve as a research subject after the paradigm of a political obligation.

The second question urges us to explore whether some individuals should be bound to serve as subjects by virtue of their meeting some specified criteria, which make them uniquely valuable to the research project, regardless of whether they have made a voluntary decision to serve based on adequate, unambiguous information about the research project.

While some ethicists have offered apparently good justification for the argument that a duty falls on the shoulders of members of a community to participate in biomedical research as human subjects, closer analysis of the arguments reveals significant gaps in reasoning that we need to address. In any given situation, we cannot definitively say which members of a society have the moral duty to serve as subjects. Even if the presence of duty is established, determining exactly what action constitutes fulfillment of that duty is difficult. The degree to which we can hold identified individuals responsible to perform designated actions considered adequate to satisfy those duties is debatable.

Some justification exists to hold that people have a moral duty to participate as research subjects, and that we can encourage people to fulfill this duty even given the standard informed consent model that allows them to refuse participation based on the disclosed information that they receive. A basis exists for the view that members of vulnerable populations have a duty to serve as subjects in research projects calculated to benefit members of their group.

In practice, strict conditions are imposed that limit recruits' suitability for inclusion as research subjects. As a result, members of some groups who are the targets of the research are often statistically under-represented in clinical trials designed to find treatments they need. These excluded individuals suffer loss of possible benefits that they otherwise might have gained from participation. We must question the validity of any results of research not conducted with the target population that the research was undertaken to help. If we want to create a situation where we allow people in vulnerable populations to exercise their moral duty to participate as research subjects, thereby also improving the quality of the research results, then we need to relax requirements imposed on potential research subjects.

We can take the reasons given for including more representatives from these groups of participants to mean that prevailing standards for protecting the interests of human subjects in biomedical research are too strict and that we should not regard informed consent for the participation of human subjects as an absolute requirement.

2. Why Ask the Question?

The Declaration of Helsinki states, "In medical research on human subjects, considerations related to the well-being of the human subject should take precedence over the interests of science and society."[1] If considerations related to the well-being of the human subject were to take precedence over the interest of science and society at all times, the individual would retain the right to decide whether to participate in research undertakings regardless of how useful or urgent such research might be.

A person who decides to participate in research as a subject is entitled to utmost protection from the possibility of harm. Contemporary discourse on informed consent reflects this idea and international guidelines pertaining to the ethics of human subject involvement in health research have enshrined it. The Nuremberg Code set the trend for this attitude by providing that "the voluntary consent of the human subject is absolutely essential" and by dictating, "[T]he human subject should be at liberty to bring the experiment to an end."[2]

If we all have a duty to serve as research subjects, we would have to be experimental participants in a biomedical study at one time or another, whether we wanted to or not. We would have the duty to assume some risks, in a research context, for the benefit of others. The decision whether to participate would not be left entirely to us because a duty is a compelling reason for a person to act regardless of personal preference. In that case, the Declaration of Helsinki would not appear to be completely correct.

We sometimes hear grumblings from the scientific community that the ethics review process is an obstacle to biomedical research that slows down scientific progress. Frustrations expressed by some scientists, research institutions, and funding agencies pressure ethics committees to liberalize review requirements. One example of how such pressures have been operating is in the proposal of the Food and Drug Administration of the United States' Department of Health and Human Services (USFDA):

> We are proposing to replace the requirement that such studies be conducted in accordance with ethical principles stated n the Declaration of Helsinki with a requirement that the studies be conducted in accordance with good clinical practice (GCP), including review and approval by an independent ethics committee (IEC).[3]

Although the USFDA claims that the intention behind this proposal is to "update the standards for . . . foreign studies and to help ensure the quality and integrity of data obtained from such studies,"[4] we doubt that the purpose can be served by removing a set of guidelines that has gained acceptance worldwide and substituting it with another set that has been widely identified with the pharmaceutical industry.

If human beings have a duty to serve as subjects of research, researchers can make a claim for more liberal consent requirements. For example, the number of people afflicted with a particular disease may be seen as being so large that to find solutions quickly enough, the interests of several people who are in a position to serve as research subjects has to be sacrificed by compelling them to participate. The perceived threat to public health and safety could be seen as so serious to justify exceptions to regulations that were formulated to ensure that human subjects are not being exploited in the course of experimentation on human beings.

By asking questions regarding duty to participate as research subjects, I take the opportunity to focus attention on issues of justice in the recruitment of subjects for biomedical research from a fresh perspective. Most current discussions converge on finding ways to ensure that those who assume the burdens also enjoy proportionate benefits. I approach the justice issue from the other side by exploring the need to ensure that those who have enjoyed benefits—or are most likely to enjoy benefits—share the burdens of research proportionately. Most guidelines are concerned with ensuring that research subjects acquire access to the fruits of their participation. Here, we ask if those who stand to gain benefits have a duty to make such benefits possible in the first place.

3. Medical Duty and Political Obligation

First, let us examine the idea that the duty to serve as research subjects derives from a more general obligation of citizens to promote the interests of their country. Can we derive the duty to serve as a research subjects from the obligation of citizens to uphold institutions and laws that are indispensable for the security and safety of society, such as those regarding payment of taxes or rendering military service?

Payment of taxes is essential for the survival of society. Governments and citizens deliberate and debate systems of taxation, which taxes to impose, and the amount of tax to collect. Ultimately, after a representative government makes its decision, everyone must obey as a matter of duty. People who do not pay must either accept the punishment for the nonperformance of duty or disengage from their society. For example, some rich individuals have moved their residences or opened bank accounts in different locations to obtain more favorable taxation arrangements.

Under conditions of grave danger, rendering military service is indispensable to the survival of a society. States have the authority to coerce citizens to provide military service to ensure that it prevails over security threats. It can punish violators or withdraw recognition of their citizenship.

The duty to serve as research subjects is arguably analogous to the duty to pay taxes or the duty to render military services. Without biomedical research, society could not proceed with its activities in a healthy way. We can argue that knowledge gained from biomedical research is essential to the survival of society, and that governments should mandate that such research continue. To ensure its continuation, some individuals ought to be obliged to serve as research subjects. If we accept these premises, can we draw the conclusion that those who refuse to serve should accept punishment or suffer banishment from the society whose laws they do not obey?

I think not. We cannot say that biomedical research and the duty to serve as research subjects are as essential to the survival of society as military service in times of danger or the need to pay taxes to finance the operation of governments. Biomedical research might be a key factor contributing immensely to the flourishing of society, but we cannot regard it as necessary for society's existence. Its importance does not rise to the level of that of taxation and military service. For this reason, we should not count the perceived duty to serve as research subjects among the imperatives legitimately enforced by countries' authority to coerce.

4. Medical Benefits Model

Even if we do not recognize the duty to serve as research subjects as a political obligation or invoke the general power of society to impose its authority upon its citizens, we could cite the privilege to enjoy medical benefits essential for society to flourish as a reason to justify human beings' duty to serve as research subjects. These medical benefits require knowledge gained from research involving the participation of human beings as experimental subjects.

Citizens should contribute to society in ways directly related to the kind of benefits they derive. Since people derive benefits of a medical kind, they are obliged to provide contributions also of a medical kind. Without like kind contributions, we cannot generate those kinds of benefits. Those benefits may not be as crucial as national survival or security but, because we regard them as essential for society to flourish, we should consider as obligatory actions necessary to bring them about.

This medical benefits model for deriving a duty to serve as research subjects generates controversies difficult to resolve. People do not hold any widely accepted belief that governments ought to coerce individuals to render human subject duty to generate desired benefits. As an alternative, we could consider a free-market model of recruitment similar to the mode used to acquire other types of goods and services considered necessary for society to flourish. Offering incentives might attract sufficient numbers of individuals who are willing to offer their services. Incentives could be cost-effective because the number of individuals needed for trials is quite small as compared with the population that we might perceive as having the duty to serve.

In many parts of the world, doctors and teachers are in short supply. Yet to even entertain the idea that members of society should be obliged to serve as doctors or teachers, however necessary they are for societies to flourish, is preposterous. One reason is that people in these professions require special training and skills before becoming qualified to practice competently in society. On the contrary, people who serve as research subjects do not ordinarily require special training or professional qualifications, although they must meet some specified criteria for each project. Inclusion criteria for subjects in research projects are highly variable among projects and so narrowly defined, unique to each project, that anchoring recruitment effort on a general duty to serve as research subjects makes no sense.

In the end, the duty to serve arises not from a general obligation to help bring about desired medical benefits but from an individual having the inclusion criteria corresponding to particular projects. If so, the medical benefits model results in a disproportionate burden to serve falling on vulnerable populations.

5. The Fairness Model

The fairness model for deriving the duty to serve as research subjects can rectify some weaknesses of the medical benefits model. According to this model, all individuals have a duty to serve as experimental subjects because they have a duty to share the burden of helping to attain medical benefits that are indispensable for any individual, and therefore society, to flourish. Overall, human beings derive medical benefits arising from knowledge gained through the conduct of biomedical research. The knowledge gained in this way would not have been possible without some human beings participating as experimental subjects. Such service requires that these individuals subject themselves to corresponding risks.

Fairness dictates that all that stand to gain benefits from biomedical research, not only a select few who are willing to serve as subjects, should participate in some relevant way to the research process. For the whole system to be fair there should be no "freeloaders." The need to be fair generates the duty to serve.

This source of the moral imperative to serve as research subjects is also problematic. First, individuals who gain benefits from the participation of others in biomedical research pay for those benefits in some form or another. Some pay directly for particular services while others pay indirectly through insurance premiums. In countries that have a good national health service, the citizens pay for that service indirectly through taxes. The claim that those who do not get involved directly as experimental subjects of biomedical research are freeloaders is incorrect.

Second, some of those to whom medical benefits are made available may not be interested in those benefits at all. Some people, for example, might be interested in nothing more than herbal remedies that do not necessarily require the involvement of human subjects in clinical research. Obligating these people to serve as subjects for research projects in which they have no interest would not be fair.

Third, in many countries throughout the world, the benefits of biomedical research are not accessible to all. Even when these are available, access is often inequitable. Where people have not experienced fair access to medical benefits, appeal to their sense of fair play by arguing that they have an obligation to help bring about the desired outcome makes no sense.

Finally, even if we can show that people whom gain medical benefits ought to contribute their fair share to help bring about the desired outcome, no necessity entails that their contribution must be in the form of participation as subjects in biomedical research. They may be able to contribute more or better to the biomedical enterprise by performing related activities, such as by being investigators, being part of Institutional Review Boards or ethics review committees, by providing funding for some research, or by serving on the staff of biomedical research institutions. They could even contribute to the effort to bring about the desired medical benefits by cleaning laboratories or lavatories.

6. Harm Prevention Model

A harm prevention model of duty to serve as experimental subjects derives from the idea that the prevention of harm is a fundamental function of medicine. We can prevent or minimize harm related to ill health through biomedical research conducted with human beings as research subjects. The participation of human beings as experimental subjects of biomedical research studies is essential to the prevention of medical harm. This reasoning forms the foundation for the conclusion that we ought to consider participation as research subjects to be a moral duty incumbent on human beings.

Peter A. Singer summarizes the harm prevention model in these words:

> If it is in our power to prevent suffering and death from lack of medical care without sacrificing anything else morally significant, we ought, morally, to do so. It is in our power to prevent suffering and death from lack of medical care by serving as research subjects for crucial biomedical research. Hence, we ought morally to serve as research subjects.[5]

Arguments such as Singer's sound persuasive until we examine the circumstances under which duty is supposed to arise. Then the many questions that arise include: Under what circumstances do we have the power to prevent death and suffering by serving as experimental subject in biomedical research? Considering the many threats that can confront us, what particular cases of suffering and death do we have a duty to prevent? In cases where persons are in a position to prevent the suffering or death of a given type of patients by their participation in a particular research project, how should individual duty be determined?

These questions illustrate that a duty to serve as research subject derived from a harm prevention model is so indeterminate that the model provides us little, if any, guidance about how to judge whether any given individual has a duty to serve in any given research project.

Even if we were to recognize a particular person as having an obligation to help prevent another's death or suffering, we require more information before we can determine if that individual should fulfill his or her personal obligation by serving as a research subject in any given study. Some individual could be better qualified to prevent the death or suffering by treating the patient directly as a doctor, by providing palliative care of some sort, or by doing something else of an entirely non-medical nature.

Samaritanism, a term coined in literature concerned with the derivation of political obligation, is similar to the harm prevention model. According to this view, people have an obligation to come to the aid of others who are in peril or dire need, and this obligation forms the justification for governments to coerce citizens into aiding other citizens. Christopher Heath Wellman explains:

> [S]amaritanism is the key to explaining how the benefits of a state justify its coercion. More specifically, the only non-paternalistic way to suppose that the advantages of political society permit a state to coerce all those within its borders is to maintain that the perils that others would experience in a state of nature can limit our own moral rights.
>
> The common understanding of samaritanism is that one has a duty to help a stranger when the latter is sufficiently imperiled and one can rescue her at no unreasonable cost to oneself.[6]

Applied to the field of health care, this view would hold that biomedical research addresses concerns of people in peril or dire need. Since biomedical research could not successfully occur without the participation of human beings as experimental subjects, human beings have a duty to serve in that capacity as long as such service does not entail unreasonable cost or harm to them.

Like the harm prevention model, samaritanism leaves us with unanswered questions about the circumstances under which a duty to serve arises. Criticism based on the variability of inclusion criteria for different types of biomedical research also retains its validity.

If we ascribe a duty to serve as a research subject to a particular individual on the model of harm prevention or samaritanism we have to justify the ascription further in terms of the match between the inclusion requirements of the particular research project and the medical condition of the prospective subject.

7. Responsibility to Future Generations and Reciprocity Models

Responsibility to future generations as the basis of duty to serve as an experimental subject of biomedical research begins with the premise that present generations benefit from the fruits of biomedical research because of the voluntary participation of past generations as research subjects. The fruits of biomedical research are indispensable for a world conducive to human survival and flourishing. As we have gained benefits from the effects of the participation of past generations in biomedical research, we now have an obligation to future generations to make similar benefits available to them through our involvement as research subjects.

The enjoyment of benefits engendered by the use of human subjects in past generations creates a kind of indebtedness owed not to those past generations but to future generations. We can say the same about the individually oriented reciprocity model of derivation.

The reciprocity model is the view that human beings who have gained benefits from the knowledge obtained from biomedical research have a duty to reciprocate the benefits received. Such research required the participation and the assumption of risks by others. Those who have benefited from the results have a duty to repay those who originally took the risks by serving as subjects of biomedical research that will benefit future generations. They direct the reciprocation forward to future beneficiaries instead of backwards to past subjects because the benefits are prospective. Having gained from the efforts of those who participated in the past, people find themselves obligated to benefit others who will have some future need.

I contend that these arguments fail to establish an inter-generational debt of gratitude because the indebtedness is generated from voluntary, not obligatory, participation by past generations in biomedical research. Since past generations were not obliged to perform the services from which present generations are now able to derive benefits, present generations do not necessarily have to agree that they are under any obligation. For the analogy to be sustained, they ought to see themselves as being in a position to render services as research subjects voluntarily as a way to benefit future generations. We cannot construe rendering services voluntarily as an obligation.

8. Conditions under which a Duty to Serve as Research Subjects May Be Derived

In general, different models for deriving the duty to serve as research subjects are vulnerable to the objection that they lack the particularity that could connect specified individuals to a moral imperative to render service as subjects of biomedical research in particular cases or situations. Still, the objections raised by critics fail to negate the importance of biomedical research for human beings or the urgency of conducting particular research projects under some circumstances. Examining conditions under which the derivation of a duty to serve as research subjects may be tested is useful.

For example, let us examine the appropriateness and usefulness of speaking of a duty to serve as research subjects in emergencies. Does it make sense to speak of a patient in an emergency as having the duty to serve as research subject?

We can hardly regard patients in emergencies as having the competence to give informed consent. Even when they are conscious and have the ability to communicate their wishes to care givers, they are under so much stress that we cannot presume those expressed wishes to be the outcome of free and informed deliberation. Often, the same condition that renders them unable to give informed consent makes the patient uniquely qualified to be an experimental subject in some research project. Research that studies emergencies can be urgently needed to benefit the patient in question. If we can derive a duty to serve as research subjects at all, the duty would be to the patients themselves. We could be talking of an experimental surgical procedure to save or prolong the patient's life, or the use of an available drug for a purpose different from that for which it has been previously approved.

To the extent that research undertakings present a reasonable likelihood of promoting the welfare of emergency patients through their participation in the research, we should regard their participation as obligatory if, in the first place, we can regard someone as having a duty to promote his or her welfare. Apart from being potentially medically useful, we can regard the procedure as having moral urgency.

Deriving a duty in this manner has little benefit because the primary importance of participation for the patients lies not in the use of the knowledge gained from the experiment but in the direct effect of their participation on their condition. The procedures entailed in their participation have greater significance for their therapeutic, instead of experimental value.

Investigators also conduct research on emergency patients with a view to discovering knowledge that could help save or prolong the lives of similarly situated patients. Notwithstanding the strong emphasis given to informed consent requirements in the ethics review of health research, exemptions have been justified in emergencies justified by the potential to benefit future patients.

For example, the United States Federal Food and Drug Administration (FDA) allows a waiver of the requirement of obtaining informed consent from research subjects or their surrogates in some cases of emergency research.[7] Critics have assailed the waiver for failing to uphold the informed consent protection deemed necessary for human subject research in general. Some saw the opportunity to give recognition not only to "the potential benefit to some patient-subjects (those in the treatment group) who receive new therapies if those therapies are successful" but also "the social need for research in an emergency to test treatments for patients presenting with acute crises such as strokes for which we have few valuable treatments that limit the resulting disabilities." [8]

Baruch Brody takes note of the need to stop "seeing the moral world as governed by the protective conception of justice" since "justice consists of balancing many independent values, none of which are absolute [and] it is this balancing conception of justice that justifies the approach found in the FDA regulations."[9] Based on this account, the emphasis on balancing protection for vulnerable subjects in research with access to the newest therapeutic agents has been overemphasized at the expense of "the rights of other patients who are not subjects in the research protocol to protection against inefficacious or dangerous drugs and to quick access to these therapeutic agents if, and as soon as, the benefits of their use have been adequately established."[10]

Effectively, this argument makes the claim that the burden lies not only on researchers to be more aggressive in their efforts but on research subjects to be more willing to face the risks incumbent in the procedures and methodologies involved. For as long as the participation of human subjects is a basic requirement for the conduct of research intended to benefit human beings in general, the moral urgency of undertaking particular research protocols translates into a moral imperative for the services of prospective experimental subjects.

For example, if an emergency patient has been determined to meet the inclusion criteria for ongoing research of urgent importance and great potential benefit for many similarly situated patients, a moral imperative arises for that particular patient to participate in some manner. Not merely a general imperative that could fall on someone else's shoulders, this duty befalls a particular patient at a particular moment in time. Though the patient fits the pertinent inclusion criteria being attributable to a chance occurrence, the patient is in the particular sort of health condition when the need arises for research subjects who fit the pertinent inclusion criteria.

A crucial requirement in the derivation of this duty to serve as research subjects is the similarity in the condition of the patient and of those who could benefit from knowledge gained from the pertinent research. This similarity binds the person being recruited to the beneficiaries of the research in a way that does not apply when dealing with patients whose conditions are not so limited or narrow. This point appears to be recognized by provisions in international guidelines that allow the involvement of members of special populations in research while discouraging their recruitment in research that could be conducted equally well on "normal" subjects or in research not explicitly intended to obtain knowledge beneficial to members of that particular population.

The link between the proposed subject and those who can gain benefits from the subject's participation is crucial. Investigators recruit these proposed experimental subjects because of the condition that renders them vulnerable. That particular condition or vulnerability connects them to those—other than themselves—who stand to profit from the knowledge we can gain through their participation in the research.

Notwithstanding some differences, we can say the same things about children and others who lack competence, people with serious and life-threatening illnesses, and pregnant or lactating women:

(1) People who belong to the category under study are in the best position to contribute to medical research in a way that other people cannot.

(2) In the event that the research is successful, people who belong to the category under study are in the best position to derive benefits.

(3) In the case of research decidedly urgent and crucial in the effort to find ways of preventing or limiting serious illnesses or conditions, those who belong to the class of individuals under study have a duty to serve as research subjects.

(4) The commonality (in terms of their vulnerabilities and the requirements for dealing with such vulnerabilities) between the prospective subjects and the study's potential beneficiaries provides a basis for generating a duty to serve as research subject.

(5) Their having the duty to serve as research subjects does not automatically translate into losing the right to render informed consent. A waiver of informed consent requirements is not justified because the patient could have other duties that are in conflict and the type of duty that arises is one that requires affirmation by the patient in terms of such conditions as deemed essential for a valid informed consent.

(6) While the commonality does not constitute a sufficient basis for overriding informed consent considerations, it adds to the evidence with which prospective subjects must reckon in deciding whether to heed the call of duty.

(7) The conditions that make them candidates for significant or urgent research also bind them to those who can gain the most from their participation. Even if they themselves were not in a position to benefit directly, those who stand to gain the most benefits are "one of them" or "people like them."

(8) Compared to others who are not similarly situated, such subjects are better able to understand and appreciate the subjective experiences of those whose health can improve because of insights drawn from the outcome of the research.

9. The Urgency of Research

We must evaluate the urgency of research at two levels: (1) a general one relating to the seriousness of the health condition for which patients are seeking remedies, and (2) a specific one relating to the necessity of a particular individual's participation. A duty to serve as research subject arises only when the research is urgent at both levels.

Urgency at the first level can be determined by society based on data and information pertaining to the damage wrought by a condition and its general cost to society. Urgency at the second level—at the level of individual subjects—has to be determined by particular individuals or of advocates acting in their behalf.

To take the example of research on children lacking competence to render informed consent, we can determine urgency at the first level by assessing the extent to which children are exposed to risks arising from the use of drugs not tested on their kind and are denied access to therapeutic advances:

> [O]nly 20 percent of drugs approved in the United States have been labeled for use in infants and children and only 37 percent of new drugs in 1996, with the potential for pediatric use, had some pediatric labeling at the time for approval. It follows that inadequate information exposes children to unexpected adverse reactions or to suboptimal treatment. In addition, the lack of pediatric formulation of a drug may deny a child access to an important therapeutic advance, or expose a child to a drug in homemade, poorly absorbed preparations.
>
> In other words, since children are not sufficiently included in research, adequate information is lacking to guide the pediatric use of a medication. Thus any particular child is exposed to greater risk, and may be deprived of potential benefit, when compared to an adult, given the same medication—an unfair difference based solely on age. This situation may be due to regulatory impediments, economic disincentives, or reluctance to require pediatric studies unless the primary use of a drug will be in children. Also, the logistics of performing pediatric research may be more difficult given, for example, the lower incidence of a disease in children, the complexity of developmental changes that may take place, and the limited market for the recapture of the costs of drug development.[11]

A moral imperative to formally study drugs in children arises because of the need for therapeutic agents that are in existence or under development to be available to them. When we fail to conduct pediatric clinical drug trials, children lose potential benefits. At the same time, researchers a valuable opportunity to collect useful data is missed.

The moral imperative is an imperative to cure an injustice that beckons to the individual subject as a duty at the second level of urgency. In this regard, Robert M. Nelson admonishes us to:

> explore through careful research the capacity and limitations of children to understand and agree to experience risks as part of a research project" and promote a just situation by supporting "a fair distribution of the burdens and benefits of research . . . through allowing the voluntary and in-

> formed assent of a child to influence the risk restrictions placed on a child's participation.[12]

We must examine the second level of urgency in relation to the conditions that link a subject to the proposed research. At this level, to convince individuals to discharge a duty to serve as research subjects, the potential subjects or their representatives need to appreciate the importance of the research project.

We can use the case of women of childbearing age to illustrate these two levels of urgency. Often, advocates invoke potential harm to fetuses as an argument against participation in research. This view exemplifies the great care with which we should regard the health conditions of women of childbearing age. The possibility of such harm is real and attaches to women in the given age range who have a reasonable likelihood of getting pregnant. While we must carefully protect these women because they are vulnerable, we must also be mindful that if, in an effort to protect them, we avoid doing biomedical research on conditions to which members of this population are vulnerable, "the health needs of pregnant women will continue to be slighted, and harms likely will occur both to pregnant women and to their offspring."[13] The recognition of such urgency constitutes part of the basis for deriving a duty for these women to serve as subjects of pertinent research.

At the same time, we must address the second level of urgency. Recognizing the plight of pregnant women in general is not sufficient. Being prospective beneficiaries of successful research, we should make women aware of the call of duty. Researchers need to inform the individual woman who stands to bear the risks and burdens of particular research about the anticipated consequences, and each women should be given an opportunity to make a well-informed, independent decision whether to participate. Each has to be able to make an informed judgment that the general urgency translates into an individual duty. This idea poses a problem for our conception of duty and responsibility.

10. Duty and Responsibility

If we can say that some persons have a particular duty, we should also be able to say that if those persons fail to perform their duty, then society should hold them morally responsible for their failure. Can we then find particular situations wherein we can hold a person morally responsible for failing to serve as subject of biomedical research? No. Even in situations where no significant risks to the research subject exist, we must allow each person to exercise autonomy in decision making.

Autonomy of individuals notwithstanding, individuals might embrace their responsibility if we help them to realize that they are the only individuals who meet the criteria to participate in a particular study, and if we explain the objectives effectively, especially if the objectives are ones with which the potential subjects identify.

Others may not have a right to hold individuals responsible, but individuals' personal commitment and feeling of connectedness could generate that corresponding feeling of responsibility.

The appeal, in cases of the nature discussed herein, is to a personal connection between the person being recruited and other human beings with whom that recruit shares common, binding characteristics. Even when no blood or close emotional ties exist between subject and prospective beneficiaries, a fundamental appeal is generated when we are able to cite bonds of a personal character. We find these bonds in shared sufferings and experiences common to those who have been afflicted by a particular disease. These bonds generate willingness to exercise a moral duty to serve as research subjects.

ACKNOWLEDGMENT

The author acknowledges the support of NIH Grant Number R25-TW 6061 from the International Bioethics Education and Career Development Award Program of the Fogarty International Center (FIC). The ideas expressed are solely the author's responsibility and are not necessarily representative of the official views of NIH.

NOTES

1. World Medical Association, "Declaration of Helsinki: Ethical Principles for Medical Research Involving Human Subjects, Edinburgh, Scotland, October 2000," in *World Medical Association Recommendations Guiding Physicians in Biomedical Research Involving Human Subjects* (Ferney-Voltaire Cedex, France: World Medical Association, 2002), par. 5.

2. *Trials of War Criminals before the Nuremberg Military Tribunals under Control Council Law*, 2:10 (Washington, D.C.: United States Government Printing Office, 1949), pp. 181–182.

3. Department of Health and Human Services, Food and Drug Administration, "Proposed Rules," Federal Register, 69:112 (10 June 2004), 21 DFR pt. 312 Docket No. 2004N-0018, pp. 32467–32475.

4. Ibid.

5. Cf. Peter A. Singer, "Famine, Affluence and Morality," *Philosophy & Public Affairs*, 1:3 (Spring 1972), pp. 231–243.

6. Christopher Heath Wellman, "Toward a Liberal Theory of Political Obligation," *Ethics*, 111:4 (July 2001), p. 744.

7. Department of Health and Human Services, Food and Drug Administration, "Protection of Human Subjects; Informed Consent and Waiver of Informed Consent Requirements in Certain Emergency Research; Final Rules," *Federal Register*, 61:192 (2 October 1996), pp. 51497–51531.

8. Baruch Brody, "Research on the Vulnerable Sick," in Jeffrey P. Kahn, Anna C. Mastroianni, and Jeremy Sugarman, eds., *Beyond Consent: Seeking Justice in Research* (New York: Oxford University Press, 1998), p. 35.

9. Ibid., p. 36.

10. Ibid. p. 38.

11. Robert M. Nelson, "Children as Research Subjects," Kahn, Mastroianni, and Sugarman, *Beyond Consent*, pp. 47–48.

12. Ibid., pp. 62–63.

13. Nancy Kass, "Gender and Research," Kahn, Mastroianni, and Sugarman, *Beyond Consent*, pp. 79–80.

Thirteen

VULNERABILITY IN BIOMEDICAL RESEARCH: A FRAMEWORK FOR ANALYSIS

Elma Lourdes Campos Pavone Zoboli

1. Two Forms of the Question

A fundamental ethical principle in biomedical research with human subjects is respect for people. This incorporates two convictions: We should treat individuals as autonomous agents, and we should protect people with reduced autonomy—vulnerable subjects—when we ask them to give consents.

To understand autonomy and vulnerability in all their complexity, we must first look beyond the relationship between researcher and subject and the obtaining of informed consent. This is because respect for the autonomy of people requires an understanding of the social conditions under which they live.[1] We must consider carefully those socio-structural factors that interfere in the life and health of people because the repercussions of these factors ultimately determine the most circumscribed dimensions of the relationships.[2]

Any proposal to contextualize and broaden the debate on vulnerability of human subjects in biomedical research must address it in its different dimensions: the individual and the collective. To do this, we will draw a parallel with the approach of vulnerability as developed in the conceptual and practical field of intervention for the pandemic of AIDS.

Based on this comparison, this chapter will present a framework for analysis of vulnerability of human subjects in biomedical research. The framework proposes the analysis of vulnerability on three interdependent levels: individual, programmatic and societal. The objective is to shift the focus of vulnerability from the individual to a collective perspective. The chapter begins with a brief outline of historical aspects of efforts to control the pandemic of AIDS to further aid in the understanding of the genesis of this approach to vulnerability. Then we see a parallel between the concept and the framework analysis of vulnerability used with AIDS and biomedical research ethics.

2. Vulnerability and AIDS: From Epidemiology to Human Rights

Since 1981, in the effort to control the pandemic of HIV/AIDS, a central theme has emerged, that HIV prevention will require efforts at the individual, community, and societal levels. Those involved in fighting this pandemic learned, after many victories and loses, that only a combination of all three would be effective.

According to Jonathan M. Mann and Daniel Tarantola, mentors of the vulnerability approach for HIV/AIDS, we can identify three periods in the history of HIV/AIDS.[3] The Period of Discovery (1981–1984) started with the recognition of the new clinical entity. Epidemiological studies provided descriptive information of the syndrome, tracked routes of pandemic spread and identified risk behaviors for HIV infection. Pandemic control efforts focused on providing information about risk behaviors to stimulate individual behavior change, with advisories to limit the number of sexual partners and practice safe sex. The following figure, as proposed by Mann and Tarantolla, illustrates the risk reduction strategy focused on the individual:

Individual	←	Risk reduction: Information/Education

The central goal of the Period of Early Response (1985–1988) was individual risk reduction. Researchers took unprecedented efforts to promote and support individual risk reduction through individual behavioral change. The support was given through services and activities in a manner similar to traditional public health responses. There was an effort to implement HIV prevention programs in countries around the world. These prevention programs' designs provided information and education along with health and social services needed to promote and support the recommended behavioral changes. These services included condom distribution, HIV testing and counseling, treatment for other sexually transmitted diseases, needle exchange, treatment for intravenous drug users, and the provision of safe blood and blood products.

An innovative element included in these programs was a requirement to prevent discrimination against HIV-infected people and people with AIDS. Discrimination was a tragic and counter-productive effect of the pandemic, which endangered public health. For the first time in history, preventing discrimination against infected people became an integral part of the strategy to control an epidemic. Mann and Tarantola illustrate this expanded risk reduction approach, focused on the individual, protection against discrimination, and access to services, in the following schematic diagram:

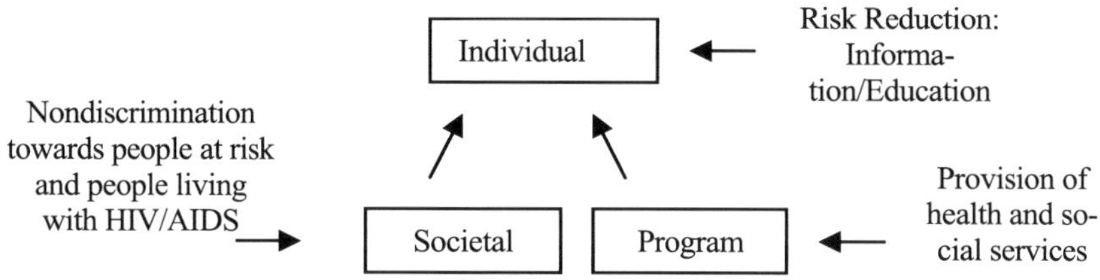

During the Current Period (1989 to present), the HIV risk reduction approach remains necessary and useful, but is not sufficient to control the pandemic. Worldwide experience demonstrates that an exclusive focus on individual risk reduction is too narrow because it fails to deal concretely with the day-to-day social realities of men, women, and children. A societal dimension extended and broadened the approach to risk reduction. The concept of vulnerability has been central in this effort. By focusing on constraints and barriers to individual control over health, vulnerability analysis can consider political, social, cultural, and economic influences on decision making, behavior, and health. We have seen a shift in focus from individual risk reduction to societal risk reduction. Mann and Tarantola offer a schematic view of this approach, which combines risk reduction strategy and efforts toward vulnerability reduction:

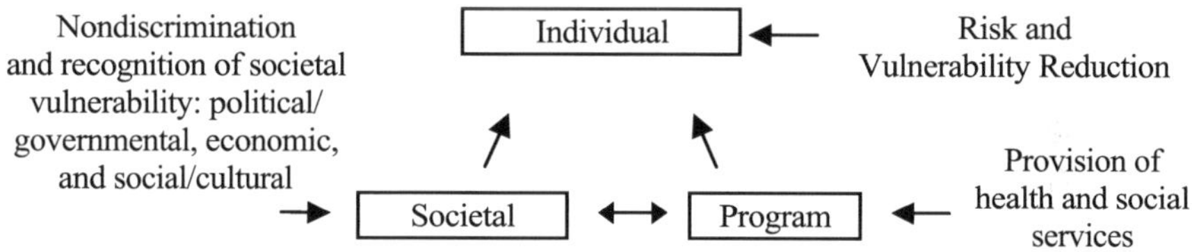

The different levels of analysis and approaches, which were typical of each period in the history of pandemic control, are all useful. Each successive strategy goes beyond previous ones.

Vulnerability is the opposite of empowerment and represents the degrees and kinds of individuals' and communities' susceptibility to infection, illness, or death from HIV/AIDS, at the same time accounting for how they are situated with regard to the integrated set of social, programmatic, and individual aspects which play figure in their capacity to respond to a pandemic.[4]

According to Mann, Tarantola, and Thomas W. Netter, "vulnerability is the converse of empowerment." By vulnerability, the authors mean "the extent to which individuals are capable of making and effecting free and informed decisions about their life."[5]

To facilitate analysis, the authors proposed that we can consider vulnerability on three interdependent levels: the personal, programmatic, and societal.[6] Personal vulnerability involves cognitive and behavioral aspects. These include information on HIV/AIDS, sexuality, and services; personal characteristics, such as emotional development, perception of risk, personal attitudes towards sex and sexuality, and history of sexual and substance abuse; and personal skills, such as ability to negotiate sexual practices, including safe sex, and skills needed for condom use.

The World Health Organization defines programmatic vulnerability, focused on the success or failure of HIV/AIDS programs to influence levels of personal vulnerability, in terms of three major prevention elements: information and education, health and social services, and nondiscrimination towards HIV-infected people and people with AIDS.

The concept of societal vulnerability is based on the insight that collective, societal factors strongly influence personal and programmatic vulnerability. It focuses on the contextual factors, which define and constrain personal and programmatic vulnerability. Vulnerability analysis shows that broader, contextual issues—such as governmental structure, gender relationships, attitudes towards sexuality, religious beliefs, and poverty—influence personal vulnerability to HIV/AIDS directly or through programs.

In this chapter, I propose that we can also address the vulnerability of biomedical research subjects by these three interdependent levels of framework analysis.

3. Vulnerability and Biomedical Research Ethics: From Consent to Context

To broaden the approach of vulnerability in biomedical research ethics, the first step must be to shift the focus toward the societal context where the protocol is conducted.

We can no longer define vulnerability in terms of a subject's age or cognitive, mental, or legal competence to give informed consent. The traditional view of consent focuses solely on individual subject vulnerability, forgetting the context in which the relationship between the researcher and the subject takes place:

Individual (Subject)	←	Vulnerability reduction: Information/ Informed Consent

Defining rules and guidelines for research ethics and the implementation of ethics committees for project approval and oversight alone are not sufficient to empower subjects. But taking these steps broadens the definition of vulnerability by incorporating societal research control mechanisms and facilitating a fair social division of the burdens of research. Even this approach is still not enough to deal practically with the reality experienced by research subjects. It does not take into account the influence of contextual aspects in the making of informed consent decisions, and it does not adequately address the issue of a fair balance of burdens and benefits from research in a significant and direct manner. A schematic representation of this expanded approach would look like this:

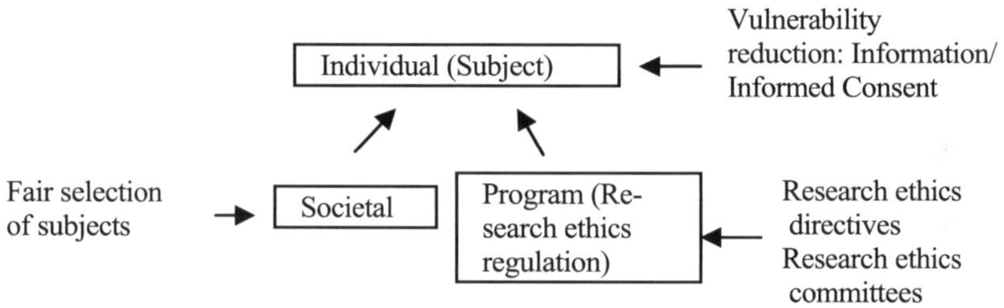

To understand vulnerability in all of its complexity, we must also incorporate considerations of constraints and barriers to subjects' control over their health, and, consequently, over their participation in or withdrawal from a study. An analytical framework for vulnerability in biomedical research ethics requires us to consider political, social, cultural, and economic influences on subjects' decision making and health. This necessitates a shift in focus from individual vulnerability reduction to a concern with societal issues. A schematic representation of this integrative approach might be:

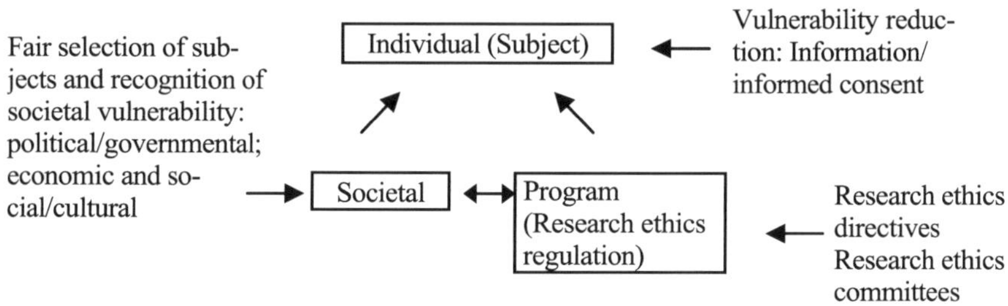

Vulnerability is the opposite of empowerment. The degree of vulnerability limits the degree to which individuals or communities are able to make free, informed decisions about whether to participate in a study. Vulnerability analysis considers an integrated set of social, programmatic, and individual aspects that play a role in a fair and balanced distribution of burdens and benefits. As in the example of AIDS control, to facilitate this analysis, we can also consider the vulnerability of subjects within the three interdependent levels framework: personal, programmatic, and societal.

Personal vulnerability is a characteristic of individuals and hinges on subjects' access to information about the research protocol, including the form in which the protocol is conveyed, which in turn influences subjects' competence to give consent. Incompetence to grant informed consent can stem from legal incapacity, as is the case with children, adolescents, or fetuses, or from mental incapacity, as is the case with adults suffering psychiatric disorders, individuals in states of unconsciousness, persistent vegetative states, or coma. Incompetence to give free, informed consent can also be caused by psychological conditioning, as in the case with those under the influence some forms of authority, for example, students, military personnel, internees, prisoners, or residents of rehabilitation centers, homeless shelters, orphanages, or religious associations.

On one hand, for informed consent to be valid, information provided to prospective subjects must meet established criteria. The consent must contain an explicit guarantee to preserve the rights of subjects to freedom, privacy, and confidentiality. These issues are comparatively easy to address in a written form. Other factors exist though, which influence and determine the validity of informed consent, but are not as easily expressed in a written form. These include the quality of the researcher/subject relationship, the understanding that the subject has about his or her rights, and the existence of alternatives for treatment or diagnosis.

Programmatic vulnerability focuses on the issues in the establishment of international and local regulations and ethical guidelines for biomedical research with human subjects, the dissemination and implementation of these directives, and the existence and the operation of Research Ethics Committees (RECs) for revision of protocols before research begins. Crucial agenda items for consideration include the logistics of forming RECs, the participation and representation of patients/subjects on these committees and in the analysis of projects, access of subjects to the committees, the publicizing of regulations and guidelines for biomedical research ethics, oversight by RECs, and independence of RECs to review and oversee the protocols.

Discussion of societal vulnerability recognizes that collective, societal factors strongly influence levels of personal and programmatic vulnerability. It focuses on these contextual factors and on conflicts of interest, which influence personal or programmatic vulnerabilities. These factors include poverty, social inequalities, access to health and education services, cultural and religious beliefs, the discrimination against and marginalization of particular groups, power relationships among social classes, gender, group leaders, communities, researchers, and subjects, and the strong influence of the pharmaceutical industry on medical services, universities, and journal publications.

For instance, in countries that have significant unmet demand for medical assistance and where access to health-care services is limited, we should ask ourselves whether a subject, after facing so much difficulty just to receive treatment, truly feels free to exercise his right to elect or reject participation. Are there other places that offer the same treatment or is this the only service available? Does the patient believe refusal to participate may mean an even greater delay in treatment? Does the patient fear that by refusing to take part his treatment team might marginalize him? Since most studies, mainly in developing countries, occur in public hospitals, we can ask, would researchers propose the same project to a patient in a private institution? Would they propose the same project in a developed country?

In an ethical analysis of biomedical research projects, societal vulnerability is crucial in situations that raise questions regarding the possibility of exploiting subjects only as a means to an end for scientific research. In these cases, subjects' competence to consent and the consent process is not as crucial as whether the proposal that these people be research subjects is just. Treating the subjects only as simple means and not as ends in themselves constitutes a violation of the principle of respect for people as autonomous agents.

The essential function of RECs is to ensure the rights of biomedical research subjects and protect vulnerable individuals. The research ethics committees might be limited in their capacity to modify societal conditions which determine subjects' vulnerability. But they can avoid some abuses, such as the use of adverse contextual conditions, as a justification for imbalances in the distribution of burdens and benefits or in subjects' recruitment.

4. A Three-Level Vulnerability Analysis Framework for Biomedical Research Ethics

Based on this multidimensional approach, we can build an analytical framework for evaluating the potential vulnerability of subjects in biomedical research from three perspectives, analogous to the framework proposed for AIDS: (1) individuals' issues of informed consent; (2) programmatic issues in the form of research ethics guidelines and RECs; and (3) societal factors that contribute to vulnerability. This approach combines prevention of individual subject vulnerability with contextual intervention. The analysis evokes the nature, strength, and interaction of determinants of vulnerability, and explores possible approaches to reduction of vulnerability on all levels. A summary of factors that we need to assess and procedural aspects of research protocols designed to minimize the potential for exploitation of vulnerable populations follows:

A. Individual or Personal Vulnerability

i. Access to Information Criteria for Valid Informed Consent

Researchers should:

Provide information in sufficient quantity and quality for prospective subjects to be able to make reasonably informed decisions about participation;
Use suitable language geared to their age and level of cognitive, psychological, and cultural understanding;
Explain necessary procedures, potential risks, and expected discomfort to prospective subjects;
Explain expected risks and benefits to prospective subjects;
Assess the presence of conflicts of interest that could result in the omission of information or other distortions in the transmission of information to subjects; and
Allow prospective subjects adequate time to weigh risks and benefits before making decisions.

ii. Competence to Decide

Researchers should assess:

The presence of factors that could interfere with or hinder subjects' decision making that include biological, legal, age, or mental factors; and values, beliefs, cultural or religious customs;
The capability of subjects to understand information provided on the project;
The capability of subjects to analyze information on the project in light of personal values and beliefs, and their state of health;
The capability of subjects to evaluate the consequences of their decisions;
The capability of subjects to predict or desire results for their health;
The capability of subjects to communicate their desires, fears, and doubts about the research or about their health; and
The capability of subjects to justify their choices in a coherent manner, according to reasonable person criteria.

iii. Personal Right to Freedom

Researchers should:

Assess the presence of the influence of authority or other conditions that might restrict the freedom of individuals to make decisions;
Assess the presence or absence of other health treatment services as alternatives to participation in the proposed research;
Ensure clauses in the informed consent form that ensure freedom of the individual to enter and leave the study;
Assure high-quality care independent of the participation of the subject in research;
Avoid the promise of greater flexibility and promptness of care as a benefit of participation in the study; and
Assess cultural and religious customs that interfere in the individual's decision-making freedom.

iv. Right to Privacy and Confidentiality

Researchers should design:

Procedures to ensure the privacy of subjects throughout all stages of research;
Procedures to ensure confidentiality of personal records of the subjects;
Procedures to ensure anonymity of subjects; and
Procedural safety measures to preserve privacy and confidentiality of subjects.

v. Quality of the Subject/Researcher Relationship

Researchers should:

Assess the presence of conflicts of interest that could result in researchers pressuring subjects to participate or interfere with subjects wishing to leave projects;
Assess power relationships, whether political or religious, that might interfere in the subject/researcher relationship;
Respect subjects as autonomous persons; and
Assess communication abilities of the researcher and subjects.

vi. Subjects' Understanding of Rights

Researchers should:

Disseminate local and international guidelines on biomedical research ethics to patients in teaching hospitals and disease associations;
Match language in the guidelines to the reading comprehension and understanding level of subjects; and
Ensure subjects' awareness of their rights.

vii. Alternatives for Treatment or Diagnosis

Researchers should:

Provide information on other medically proven treatments or diagnosis opportunities beyond that offered by the study; and
Ensure access of subjects to alternative treatments or diagnoses.

B. Programmatic Vulnerability

i. Rules for Ethics in Biomedical Research

Researchers and countries should:

Research, remain aware of, and follow local, national, and international rules for ethics in biomedical research;
Support and provide incentives for biomedical research institutions to encourage adherence to local and international biomedical research ethics directives;
Encourage participation of researchers, subjects and other interested parties in the creation and revision of guidelines for ethics in biomedical research; and
Require protocol review by an REC before allowing research to begin.

ii. Research Ethics Committees

Researchers should:

Establish research ethics committees in institutions that conduct studies with human subjects;
Learn and cooperate with the organization and functions of research ethics committees;

Ensure that RECs encourage the egalitarian participation of both genders and members of diverse segments of society;
Ensure that RECs include representatives from different professions in the biomedical and liberal arts areas;
Establish ways for subjects to contact RECs for consultations or complaints;
Ensure transparency in the dialogue and relationships of the research ethics committees, researchers, and research sponsors;
Remain mindful of conflicts of interest that might interfere with the independence needed by the RECs in the analysis of protocols;
Lobby for institutional and governmental support for the proper functioning of RECs;
Ensure that the work done by research ethics committees in the analysis of protocols and protection of research subjects is subject to oversight by society and the government; and
Ensure participation of subjects in the RECs for the analysis and oversight of protocols.

C. Societal Vulnerability Factors

i. Livelihood

Researchers should evaluate and work to meliorate negative impacts caused by:

Poverty;
Lack of adequate educational opportunities;
Levels of educational achievement;
Patterns of distribution of wealth; economic gaps within society;
Organization and quality of health-care systems;
Guarantees or barriers to health service access;
Cultural and religious customs;
Social class inequalities; and the
Power structure of local leadership with subjects and researchers.

ii. Environment Characteristics

To create a supportive societal environment, researchers should:

Promote a collective, positive view of biomedical research;
Work to avoid or reduce marginalization of groups;
Lobby for enactment of policies and laws to protect biomedical research subjects;

Support organizations that protect the rights of research subjects or patients in their dealings with RECs;
Ensure access of subjects to research benefits after studies end; and
Return benefits of the research to the community in which they conducted the research.

iii. Conflicts of Interest

Researchers should be mindful of, and work to avoid pressure created by:

Expectations to publish in periodicals that only accept research that reports positive results;
Research sponsors that attempt to influence publication of results;
The strong presence of the pharmaceutical industry in the funding of research projects, periodicals, scientific events, and researchers;
The pharmaceutical industry that funds universities and teaching hospitals for treatment, education, and research activities; and
Universities urging researchers to produce and publish.

The aim of this proposal is not to exhaustively cover all possible aspects of vulnerability. Instead, this outline is intended to serve as a model that must be refined and adapted to each situation to facilitate:

Definition of the population subset and the setting (social, cultural, economic environment) in which research is designed and conducted;
Identification of particular subject vulnerability factors in any of the categories (individual, programmatic or societal);
Determining the way that the chosen vulnerability factors might influence subjects' choices or the fair balance of research burdens and benefits;
The explanation of how the chosen vulnerability factor is influenced by or influences other factors listed within the same or other categories; and a
Course of action to reduce vulnerability on each of the three levels.

5. Conclusion

The purpose of this multidimensional analysis of vulnerability is to strengthen our ability to perceive and condemn those unjust social inequalities and conditions that impinge on individuals' autonomy. Its adoption in biomedical research ethics requires us to act collectively in building intervention strategies that address issues beyond the scope of researcher/ subject relationships.

NOTES

1. Hupert Lepargneur, "*Bioética, poder e Injustiça: uma Introdução*," ("Bioethics, Power, and Injustice: An Introduction"), *Bioética: Alguns Desafios*, eds. Christian de Paul de Barchifontaine and Leo Pessini (*Bioethics: Some Issues*) (São Paulo: Loyola, Centro Universitário São Camilo, 2001), pp. 49–63.

2. Marcio Fabri dos Anjos, "Bioética nas Desigualdades Sociais," ("Bioethics in Social Inequalities"), *A Bioética no Século XXI* (*Bioethics in the 21st Century*), eds. Volnei Garrafa and Sergio Ibiapina Ferreira Costa (Brasília, DF: UnB, 2000), pp. 49–65.

3. Jonathan M. Mann and Daniel Tarantola, *AIDS in the World II: Global Dimensions, Social Roots, and Responses* (New York and Oxford: Oxford University Press, 1996).

4. J. R. Ayres, I. França Jr, G. Calazans, and H. Saletti, "*Vulnerabilidade e Prevenção em Tempos de AIDS*," ("Vulnerability and Prevention in the Age of AIDS"), *Sexualidades pelo Avesso: Direitos, Identidades e Poder* (*Sexuality by the Reverse: Rights, Identies, and Power*), eds. Regina Maria Barbosa, and Richard Guy Parker (Rio de Janeiro and São Paulo: IMS-UERJ, 1999), pp. 49–72.

5. Jonathan M. Mann, Daniel Tarantola, and Thomas W. Netter, eds., *AIDS in the World* (Cambridge, Mass.: Harvard University Press, 1992), pp. 577–602.

6. Ibid.

Fourteen

THE NEW VULNERABILITIES RAISED BY BIOMEDICAL RESEARCH

Maria do Céu Patrão Neves

1. Introduction

The ethical question of biomedical research on human beings for therapeutic purposes is one of the first crucial and most discussed issues in bioethics.[1] Researchers whose work involves human experimentation know that they will have to present their research objectives and the respective processes involved to ethics committees for scrutiny. In this chapter, I will examine factors that triggered this now common situation and the stages of its development. I will also discuss the current proliferation of new vulnerabilities brought about by new biomedical powers, and I propose paths for reflection and action to meet the present challenges.

2. From the Necessity for Human Experimentation to the Pressing Need for Ethical Reflection

The experimental method began to be applied to the study of man in the Renaissance period. Yet only in the nineteenth century did human experimentation became indispensable for the progress of science, for the improvement of clinical practice and for modern medicine, which evolved from art to science and is now inexorably experimental.

In the aftermath of World War II, ethical reflection on human experimentation led us to the recognition that knowledge is not an absolute value and that progress does not necessarily lead to good. Science should not be the sole creator of its own design. Instead, the interests of science must be subordinate to those of society. We need to establish ethical guidelines and policy regulations for scientific development and research.

A. From Discovery of Experimental Method to Human Experimentation

The notion of experimentation arose in the Renaissance and was further developed and refined during the period of modern rationalism. A more strict definition evolved within the spirit of contemporary positivism.

During the first phase, experimentation consisted of the appreciation of facts, in observation, in the verification of ideas, and in questioning or refuting some orthodox truths. In short, experimentation related experience to reason.

The second phase involved building the experimental method in four defining steps: observation, formulation of hypotheses, verification of laws, and determination of laws. Science no longer coincided with philosophy. It no longer consisted of thought about reality but it became empirical reality translated into thought. The third phase was the expansion of experimentation to all reality, resulting in the multiplication and development of sciences. Medical experimentation was born, and focused, systematic, and controlled experimentation on healthy human beings replaced the observation of corpses and other forms of clinical experimentation that characterized the earlier phases.

The third is the level of experimentation on human beings in the strict sense. It has arisen because of the irrepressible evolution of experimentation, justified and vindicated by the acquisition of the status of science, which has become associated with the endeavor to promote the good of man. We can define the specificity or the essential character of human experimentation in scientific and human terms. From the scientific point of view, the specificity of experimentation stems from three defining characteristics: the studies have a random design, the evaluation is double blind, and investigators compare the results with the effects of placebo substances or measures, or, preferably, to the effects of the drug commonly or previously used in the situation in question. From the human point of view, the specificity stems from the difficulty of maintaining the subject status of human beings who become objects of experimentation.

Because human beings have intrinsic, unconditional value, respect for human dignity dictates that we must not allow research to reduce human subjects to mere objects. While continuing to promote experimentation on human beings as an excellent means of obtaining benefits for humanity—to enhance the common good—we are challenged to preserve the dignity of each human subject involved in research—to preserve the individual good.

B. Phases of Evolution in Human Experimentation

We can divide the analysis of the recent history of human experimentation into two main periods of development, each of which comprises two stages. The first phase of experimentation on human beings, spanning the mid-nineteenth century through the end of World War II, a phase of discovery and implementation, saw the emergence of benefits entailed by the application of the experimental method to man, and the implementation of innovations into practice.

We can subdivide this century of progress into two sub-stages, which I call naïve innocence and uncurbed enthusiasm. Naïve innocence characterized the origins of human experimentation. Researchers worked alone, personally financed the research, and the subjects of experimentation were individual patients that the researchers aimed to treat. The researchers respected the deontological rules of beneficence and nonmaleficence, which in the era of pre-scientific medicine were not strictly applied. The doctor-researchers sought to promote the well-being of individuals without being able to predict the exact consequences of their experimental actions. The stage of uncurbed enthusiasm built on the still shaky knowledge achieved during the stage of naïve innocence. Researchers began to seek collaboration and to obtain a small degree of financial support. The number of experimental subjects increased, and the subjects were often anonymous. During this time, the main goal of researchers was the acquisition of knowledge. As experimentation resulted in more knowledge and researchers' enthusiasm increased, their awareness of the interests of the individual diminished, overshadowed by the so-called greater good of science or knowledge.

From "experimentation in trust" (to paraphrase Edmund Pellegrino and David C. Thomasma),[2] confined to the doctor-patient relationship and practiced on the researchers and their families, there was a move to "mass experimentation," applied to prisoners, orphans, mental patients, the aged, and soldiers. There was a growing conviction that the mass approach is justified by the absolute value of knowledge and the intrinsic goodness of progress.

Following these two phases, we see a clear transition from valuing the good of the individual to a concern for the common good. To cite an extreme example of this change in focus, we can consider the testimony of Nazi doctors on trial at Nuremberg. They claimed that they had not violated the Hippocratic principle of beneficence because they had always acted in pursuit of the common good of the German people.

In reaction to horrors imposed on human individuals in the name of the common good, the second period of human experimentation, a period of reassessment and regulation, saw a reassessment of the risks and benefits of experimentation and of the regulation of its practice. This period began after World War II, and it has continued to the present, with some transformations, which divide the period into two stages: a protectionist stage and an apparently contrary vindicatory stage.

During the protectionist stage, experimentation on human beings continued to develop, furthered by teams of researchers frequently financed by the state. At the same time, we saw a proliferation of ethical norms and bioethical institutions that assess the ethical legitimacy of projects and safeguard the interests of the subjects of experimentation. Experimentation on human beings continued, but society imposed legal regulations on its procedures, developed ethical standards governing its practice, and controlled its financing. To prevent the violation of human dignity, society took steps to respect the primacy of the individual and the individual good over common good.

Now another stage is evolving, and together with the first, it defines the present situation. I have called it vindicatory to emphasize the uniqueness of the present movement in its consideration of minorities previously excluded from clinical trials, who now demand inclusion, and common citizens, healthy or sick, who claim the right to act as subjects of experimentation. Biomedical experimentation continues, increasingly performed by multidisciplinary teams of researchers. The projects have ever-increasing private financial backing, especially from multinational pharmaceutical companies with an unquestioning demand for lucrative returns that outweighs other ends. Pressure to safeguard the well-being of individuals and the constant need to update ethical norms has shaped new guidelines. Unfortunately, these guidelines have become increasingly legal and political in nature, and are at imminent risk of losing sight of their original ethical justification. This is the stage of experimentation *à la carte*, or on individual demand, developed alongside supervised experimentation, in which scientists select human subjects.

In its loss of naivety and in its creation of protective measures, the vindicatory stage bears witness to our efforts to surpass the negative side effects of experimentation.

C. Ethical Reflection upon Human Experimentation: From the Advent of Bioethics to the New Challenges

In the nineteenth century, a patient, confined in a charity hospital, was poor, ignorant, and profoundly vulnerable to any suggestion made by a prestigious, wise doctor. In the first half of the twentieth century, we confined abandoned orphans, mental patients, senile elderly patients, prisoners, and soldiers in institutions, bereft of their individual interests for the sake of the common good, vulnerable to those upon whom they depended. Since World War II, the individual, healthy or sick, who has agreed to participate in clinical trials, has become vulnerable to the professional standing of the researcher and the quality of the investigation.

Bioethics originated within a North American geo-cultural context, spurred by human dramas caused by experimentation. It has since been the protagonist of a new, true humanism for the "technological age" (to paraphrase Hans Jonas),[3] striving to guarantee the respect for the dignity and integrity of the subject of experimentation. The goal has been to protect vulnerable populations in two ways, by defending (negative sense) and by promoting (positive sense) individual rationality, liberty, and autonomy, through the enforced obligation to obtain informed consent.

Today, researchers in the field of human experimentation extensively promote autonomy, thanks to the numerous regulations in bioethics, biolaw, and biopolitics for the protection of individual self-determination and of individuals with diminished competence. Nevertheless, I believe vulnerabilities are multiplying without our noticing. Since this happens under the cloak of presumed autonomy, the process has become increasingly difficult to detect. Consequently, it has become increasingly difficult to protect individuals. These perplexities define our challenge for the future.

3. A Reassessment of Vulnerability in the Context of Human Experimentation

A. The Notion of Vulnerability

Vulnerability as a term has recently entered our lexicon of common morality. Texts addressing biomedical ethics frequently use the term. Of Latin origin, the stem *vulnus* means wound. Vulnerability expresses a susceptibility to being wounded, implying a fragility of the being that the term qualifies.

Emmanuel Lévinas and Jonas were the first to draw philosophical attention to the notion of vulnerability in the 1970s. Only later, in the 1990s, did the notion of vulnerability become common in bioethical thought.

In *L'Humanisme de l'Autre Homme* (*Humanism of the Other*), Emmanuel Lévinas defines vulnerability as subjectivity.[4] For Lévinas, alterity, or the other, always comes before subjectivity, or before the I. Therefore the I, depending on the other, is always in relation to the other and this relation implies vulnerability: "The I, from head to foot, right to the marrow, is vulnerable."[5] In this way, vulnerability entered the vocabulary of philosophy as an intrinsic characteristic of human beings, a condition of humanity.

Jonas, in *Das Prinzip Verantwortung* (*The Imperative of Responsibility*), also draws attention to the relevance of the philosophical meaning of "vulnerability," broadening its reality to the whole of nature and specifying its meaning as the perishable character of that which exists. Humankind is not only perishable and therefore vulnerable, but its members also have the power to harm other beings, including other humans, in their vulnerability, and so it becomes a duty, implied by power, to answer for the vulnerability of others.

Having established the relation between "power" and "duty," Jonas' vulnerability gains a positive ethical meaning, that is, it determines an effective obligation: that of defending and protecting, caring for and taking responsibility for those who are vulnerable. In Jonas, "vulnerability" is basically "concern, recognized as duty," it is responsibility before a vulnerability which, when threatened, becomes the object of care.[6]

This aspect is especially important insofar as it shows how vulnerability, as a condition of our humanity, acquires a different nature in each of the two philosophical universes: imminently descriptive in Lévinas and structurally prescriptive in Jonas. Jonas' vulnerability has a normative dimension that guarantees its necessary operative character in the ambit of an ethic applied to biomedicine.

The notion of "vulnerability," recently introduced into bioethical reflection, is understood in its broad and general sense of a universal condition of humanity, drawing attention to our all being vulnerable beings, and, in its restricted and operative or normative sense, to the duty to care for those threatened by power. In the specific field of bioethics, that power belongs to biomedicine and is accentuated in the realm of human experimentation.

Within this context, the principle of vulnerability prescribes the duty to protect people in relation to the possible threats of biomedicine. "Vulnerability" is understood in its interrelation with "power" and "duty," in an indissoluble combination.

The specificity of the bioethical acceptance of "vulnerability" lies in its being understood in two levels—broad and restricted—and in the interrelation between "power" and "duty." Vulnerability is identified in this point of intersection and it demands protection.

B. Power and Vulnerability: The "New Powers" and the "New Vulnerabilities"

Biomedicine has emerged with therapeutic, economic, and social powers during this present vindicatory phase of human experimentation. Of these, only therapeutic power, which emerged when human experimentation became scientific, corresponds to the original aim of human experimentation. Therapeutic power lies in the ability to obtain precise knowledge and to produce effective means for the cure or control of illnesses. Bearing in mind its recognizably good purpose, therapeutic power is ethically legitimate and should be developed through the strict fulfillment of the principle of beneficence (the obligation to do good), combined with that of nonmaleficence (the obligation to avoid harm), and in the subordination of that power to the interest of man.

The increase of therapeutic power results in part from and at the same time requires a growing number of subjects for experimentation. The traditional recruitment from amongst the patients by a particular doctor, service, or institution, can no longer meet the demand—and, besides, it raises delicate ethical questions concerning the dependence of the patient-subject upon the doctor-researcher.

As a result, attention has turned toward encouraging the participation of volunteers. Bioethicists have advanced the idea, with limited success, that a moral obligation exists for individuals to serve as subjects for experimentation for the sake of the scientific and social good from which each of us may benefit in our lives. At the level of applied research, conditions make participation appealing. Participants often enjoy easy access to the best health services available, which is a valuable asset for the economically underprivileged sectors of society and for people who suffer from ill health and frequently seek health care. In the final stage of clinical trials, the therapeutic aspects of the experimental procedure may be superior to those of standard procedures, which is an invaluable benefit for chronically or terminally ill patients.

A paradigmatic case of the new vulnerabilities arising from new therapeutic powers is that of Jesse Gelsinger, in the United States, in 1999. Gelsinger, an eighteen-year-old male, suffered from a mild form of Ornithene Transcarbamylase Deficiency (OTC), a rare liver disease caused by a genetic defect. Most newborns suffering from OTC die within hours after birth, but Gelsinger's symptoms were successfully controlled by drugs and diet.

Gelsinger participated voluntarily in a gene therapy trial, which was an extremely promising technique investigated since the 1980s, but which failed to live up to expectations. The innovative trial aimed to inject a vector—a modified cold virus containing the missing gene—into the blood stream, despite previous knowledge that the vector, toxic administered in high dosages, had caused the death of baboons during animal experimentation. Gelsinger's trial was the first case in which gene therapy was tested on relatively healthy people. The National Institute of Health (NIH) and the Federal Food and Drug Administration (FDA) approved the trial although some geneticists considered it too risky for use in human beings, and the researchers responsible for the study did follow the approved protocol. But Jesse Gelsinger died four days after receiving the vector. That trial, along with the entire North-American program for gene therapy, was immediately suspended pending investigation.

To think that voluntary experimentation subjects, who are administered what we believe is a benefit after granting informed consent, are not vulnerable, is a mistake. Researchers' optimism made Gelsinger vulnerable. The expectations of alternative therapies, the availability of free health services, or of swift and free access to them, make people vulnerable. As therapeutic power progresses and offers greater benefits, it also creates conditions for the appearance of new vulnerabilities, when, reasonably or not, it makes promises that might not be fulfilled. These new vulnerabilities emerge at the international, social, and especially the individual level.

At the international level, developed countries create new vulnerabilities when they export therapeutic power without considering the specific conditions of its application in underdeveloped countries. For example, the prevention of the transmission of HIV by discouraging breastfeeding exposes the newborn to other possibly more predictable and fatal dangers. At the social level, we create new vulnerabilities when a population is subjected to diagnostic methods when no effective corresponding treatment exists. Genetic screening (for example, testing for falciform anemia in Afro-Americans) and the creation of stocks of DNA categorized by ethnic group exposes populations to potential new sources of discrimination.

At the individual level, patients for whom standard treatment offers no help vindicate therapeutic power from any source that promises hope—from fortunetellers to alternative medicine to the clinical trials underway, which sometimes recruit subjects via the Internet. The patients look for treatment for any ill from which they suffer—from hypochondria to terminal diseases that cause them to wish for a painless death. Sometimes, desperate and without other acceptable sources of comfort, they might resort to renting a room in which they will practice euthanasia. At the individual level, therapeutic power can become overwhelming and aggravate human vulnerabilities that include fear of illness, pain and suffering, death, and principally, lack of hope.

Evoking the principle of beneficence is not sufficient to morally legitimize the therapeutic power of biomedicine. Beneficence fails to protect humankind from the fascination that such power holds and which makes human beings vulnerable.

The economic power of biomedicine, unlike therapeutic power, has only recently emerged and immediately raises suspicions about its ethical acceptability. This economic power arises from the evolution of sources of financial support for research—personal, public, and later private—which has resulted in biomedical research becoming a financially viable activity.

Economic power is ethically legitimate and even praiseworthy if it complies with the principle of justice in the equal treatment of all. Today biomedical research requires heavy investments that surpass the possible financing capabilities of most gross national products.

As a result, we have seen a trend toward privatization in modern research. Biomedical research has transitioned from being conducted in universities, laboratories, and other state centers to being conducted in mega-companies, which have been created from merging pharmaceutical and biotechnological companies. The process has significant implications at the individual level. These include the growing anonymity of researchers and a consequent erosion of responsibility, and the involvement of scientists in the financial sphere, until recently unheard of. At the collective level, society has suffered a growing loss of control over the pursuits of science.

Because the economic powers financing research require profit to continue to finance more research, sponsors typically direct investigations toward predictably profitable paths. They can hold a reasonable expectation of profit only if they direct research to the needs of the richest countries and richest people. Powerful marketing campaigns support this course of action, which sometimes creates the need for the goods produced to guarantee their consumption.

In this context, Craig Venter's Celera Genomics program is paradigmatic at multiple levels. It illustrates the progressive control of private interests over public activities powered by profit, which frequently develop on the fringe of the most urgent and global human needs. At the same time the development creates new human vulnerabilities by commodifying the human genome.

Researchers have been mapping and sequencing the human genome in the United States since 1987. The genome project had a predicted span of twenty-five years, with public financing from the Department of Energy and the NIH. Over time, the project extended to different countries, and the Human Genome Organization (HUGO) was put in charge of coordinating the international project. Meanwhile, some private enterprises sought to support the project, especially Craig Venter's Celera Genomics that, in 1998, claimed it would sequence the human genome within three years, ahead of all other competitors. At the same time, Venter announced that his company would not immediately make public the sequences they identified.

The interest of private enterprises in the human genome mapping research extraordinarily accelerated the project. At the same time, because the private financial sponsors worked to retain exclusivity on the intellectual property in the interest of making sizable profits from the patenting of DNA and its subsequent commercialization, the sponsors erected barriers to the free circulation of information. Their actions undermined the original aim of the project, which was to improve understanding of the universal genetic heritage of humanity.

Despite polemics having arisen from this issue, in 1998, the European Union adopted a directive that permits patenting in the field of biotechnology as long as the genes or gene sequences have an industrial application and use of the technology does not threaten human dignity. This was a compromise between ethical demands and economic requirements so as not to deter investment in biotechnological industries and allowing research in this field to continue.

This is not to say that economically powerful sponsors of biomedical research contemplate only crucial research or that they represent an elite pursuing financial gain, or that they lessen human needs. Presently we are experiencing a "medicalization" of human life. We seek to find a response to every human ill in health services—from simple discontent by resorting to a drug, to particular limitations such as infertility by turning to medically assisted fertilization. This state of affairs bears witness to the economic power wielded at the level of the individual. Accentuating needs increases vulnerabilities and economic power develops from the vulnerabilities it creates: from the wish for total health and well-being to ambition for absolute perfection and the desire for beauty.

Economic power also influences states of affairs at the social and international level when the means to combat disease exist but become financially inaccessible. When a community or population is unable to obtain primary health care, such as vaccination or the treatment of those infected with HIV, due to financial limitations, we consider them more vulnerable than if the means of treatment did not exist at all.

To invoke the plural principle of justice—from the libertarian model (advocates rights to social and economic liberty) to the equalitarian model (advocates equal access to the good in life that the common person values)—is not sufficient to morally legitimize the economic power of biomedicine. That principle fails to protect persons from the economic power's need for profit, and this makes them vulnerable.

Social power is the most recent of biomedical powers, evolving from therapeutic and economic power, and reinforcing them. Subtle, though no less influential, social power derives from the ability to gain people's trust from the success it achieves and to mobilize society by pursuing the future goals it claims it can reach. This power, by involving society in the objectives of science, is ethically legitimate and recommendable. It should protect individual autonomy in the sense that it should safeguard the primacy of human well-being against the exclusive interest of society or of science.

Social power sometimes extends beyond the boundaries in which its application is beneficial. The media often exaggerates achievements in the field of biomedicine, causing unrealistic expectations among the public. Bearing in mind that biomedicine benefits from publicity and that this publicity is as widespread as the predictable impact of the news, we should not be surprised that the publicity is often sensationalistic. This void sensationalism is always misleading, promotes the false belief that nothing is impossible for biomedicine, and creates a myth of life free from suffering, of absolute health, or genetic perfection. The sensationalism *per se* exploits another aspect of human vulnerability, the imagination of each individual.

A case in point is a commercial advertisement that employed a computer-enhanced image of Christopher Reeve, the paraplegic American actor who portrayed the mythical superman in the cinema, walking. Televised during the American football Superbowl in 2000, Reeve appeared to get up from his chair and approach a stage where he joined other beneficiaries of biomedical research. The impact of the image was so great that many paralyzed people who viewed the image believed that Reeve had been cured. They envisaged a cure for themselves.

Christopher Reeve became paraplegic after a horse riding accident in 1995. After that, Reeve initiated the lifestyle he maintained until his death in 2004: giving lectures, participating in academic sessions, sports programs, and television talk shows, writing books, and making public service advertisements. His goals were to promote the quality of life of physically disabled individuals and to raise funds to finance spinal cord injury research. The image of Reeve walking was intended to spur the imagination of people who, inspired by the hope that such a recovery is possible, would then donate to the cause.

Social power also generates new vulnerabilities at the social, individual, and international levels. On the international level, social power is operative in the classification of regions of the world according to their unfavorable general or specific health conditions. Where health risks are considered high, officials advise against travel to these regions. For example, the bird flu crisis has caused huge economic problems, especially in Asian countries.

Exaggerated claims about what biomedicine can accomplish causes individuals to hope that biomedicine will satisfy their every need and desire, and results in the conviction that biomedical action is always beneficial. For example, the Iranian conjoined twins, Laleh and Ladan Bijani, were the subjects of the first-ever attempt to separate adult craniopagus twins. They died on 8 July 2003 very shortly after unsuccessful surgery in Singapore. While many people applauded the medical team's efforts, others raised questions about whether or not the procedure should have been attempted. Can we not ask if these twins were victims of the social power of biomedicine?

The collective level is where the effects of this power are most visible in the growing demand for the benefits promised by biomedicine. Health associations acting as lobbyists, in the syndicalization of illness, favor those who have the greatest vindicatory power and aggravate the vulnerability of the rest. The lobby for the HIV patients is far more powerful than that for the manifoldly more numerous malaria patients; so the bulk of financing evidently goes to the first group.

To evoke the principle of autonomy is not enough to legitimize the social power of biomedicine because it does not protect people from the illusions it creates and that make them vulnerable.

4. Conclusions: The Shortcomings of an "Ethic of Rights" and the Vindication of an "Ethics of Duty"

Biomedical research on human beings for therapeutic ends has always been motivated by some good. The new powers achieved by biomedicine also contribute in diverse ways to some good. Traditionally, beneficence aimed at individual good, but it did not avoid the violation of human dignity and integrity that the promotion of autonomy sought to ensure, in the preservation of the interests of the individual.

Today, in the globalized world in which new powers are developing, the good of the individual can no longer be sustained aside the common good, nor does autonomy appear to be capable of preventing the proliferation of new vulnerabilities. Autonomy tends to leave individuals to their heightened individualism, removing responsibility from the other, hidden behind the legalism of the fulfillment of established norms. But a person is a being in relation and it is in the relation itself that the vulnerability of each individual unveils and the care of the other is put into practice.

We must move away from an individualist perspective, structured by the philosophy of rights that has determined the hegemony of autonomy in the obligation to respect individual integrity, to a perspective of relation, to be structured by a philosophy of duties built on the obligation to protect the vulnerability of all by adopting attitudes of care and responsibility. The voice of the power of each individual expresses the individualist perspective; the perspective of relation is expressed in terms of one's duty in relation to others.

Today, therapeutic, economic, and social powers create new vulnerabilities. We need to reduce the effects of these powers. We can reduce these effects by complementing the traditional ethical principles with a principle of vulnerability. By enforcing an obligation to protect vulnerable persons, this will move us decisively from a logic of power to a logic of duty, and help our essential ethical relatedness to become fulfilled.

ACKNOWLEDGMENT

My thanks are due to Kathleen Calado for checking the language of this chapter.

NOTES

1. Albert R. Jonsen, Robert M. Veatch, and LeRoy Walters, *Source Book in Bioethics: A Documentary History* (Washington, D.C.: Georgetown University Press, 1998).
2. Edmund Pellegrino and David C. Thomasma, *For the Patient's Good* (New York: Oxford University Press, 1988).
3. Hans Jonas, *Das Prinzip Verantwortung: Versuch einer Ethik für die Technologische Zivilisation* (*Principle Responsibility: An Attempt at an Ethics for a Technologcial Civilization*) (Frankfurt am Main, Germany: Surkamp, 1979).
4. Emmanuel Lévinas, *L'Humanisme de l'Autre Homme* (*Humanism of the Other Man*) (Montpellier, France: Fata Morgana, 1972).
5. Ibid., p. 104.
6. Jonas, *Das Prinzip Verantwortung*, p. 301.
7. Jacob Dahl Rendtorf and Peter Kemp, eds., *Basic Ethical Principles in European Bioethics and Biolaw* (Copenhagen/Barcelona: Center for Ethics and Law/Borja de Bioètica, 2000).

Fifteen

HOW TO SAVE THE WORLD: ALTERNATIVES TO BIOMEDICAL RESEARCH

Frank J. Leavitt

1. Introduction: A Pound of Prevention or an Ounce of Cure?

The title of this paper is admittedly grandiose. Stepping down from our scholarly or scientific high horses to dream is sometimes a good thing because sometimes dreams have a way of coming true. This paper is a sequel to another paper of mine, in which I arrived at a quite pessimistic conclusion about the potential for bio-medical research to make any progress while respecting the humanity of research subjects.[1] In this paper, I will suggest some positive alternatives. Criticizing the work of other people is relatively easy. So the other paper was quite neat and orderly in its style. It is harder to propose something better. So this paper may appear a bit more messy. I beg the reader's patience.

The ethicality of medical research has improved since Nazi experimentation and the Tuskegee trials. Still, teaching a course on research ethics without its turning into a course on scandals remains a challenge. For some studies, parents volunteer their babies for experiments investigating treatments for which the babies have no medical need. The concept of autonomy only can be applied metaphorically to young children. Studies show that geriatric patients and patients in general lack sufficient reasoning for their consent to be considered informed.[2] The Hastings Fluoridation Experiment in New Zealand had the word "experiment" dropped from its title and replaced by "study" because "the locals objected to being experimented on."[3] Patients are said to have given "informed consent" for a clinical trial when they are not even told that the drug company has paid their physicians for prescribing the drug.[4] We cannot claim that a patient has been "informed" unless he or she is aware of whether any conflict of interest may have influenced the doctor's decision to prescribe this drug over another.

Although patients sign informed consent forms, physicians who sign contracts for clinical trials can be required to agree not to disclose any results of the trials without permission from the drug company. This practice is not universal, but is quite common, as I have verified by examining the relevant clauses of clinical trial contracts supplied to me (with identifying data stripped to ensure anonymity of the subjects) by the Chairman of the Institutional Review Board of Soroka Medical Centre, our Faculty's teaching hospital.

Those who, despite drug company refusal to grant permission, feel bound by conscience to disclose adverse affects may be persecuted for years.[5] A well-known example is that of Nancy Olivieri, of the Hospital for Sick Children in Toronto, Canada. As verified by a commission of the Canadian Association of University Teachers (CAUT), Oliveiri violated the non-disclosure clause of her contract with Apotex after concluding that the drug she was testing was putting her patients in danger. According to the CAUT report, after Apotex failed to give Oliveiri approval to publish, she informed her patients of her conclusions, which she also published in an article in the *New England Journal of Medicine*. She was sued by Apotex, demoted from her position at the hospital, and received hate emails from a colleague who sided with the drug company.

When we step back and look at the situation from a perspective of detached concern, we realize that we are in a dilemma. On the one hand, we need medical research. On the other hand, requirements based on bioethics have become so strict carrying out medical research ethically appears impossible.

How can we make any progress in pediatric medicine unless we adopt the fiction that surrogate consent granted by the parents, even in non-therapeutic research, constitutes valid informed consent? How can we do any research on patients unless, despite the possibility that their illness impairs their decision-making capability, we pretend that their consent is genuinely free and informed? How can drug companies recruit physicians to conduct clinical trials without rewarding them for their efforts? What patients will accept experimental drugs if they are informed that their doctor is receiving money or a free trip to a conference on some beautiful island in return for prescribing them? What drug company will finance expensive trials unless they have some control over the results?

One solution to the dilemma would be to admit that much or most medical research cannot be conducted ethically. We can then choose either to drop ethics or to drop much or most of the research. In fact, the second choice has largely become the norm. Despite the inestimable benefit that Edward Jenner's smallpox vaccine experiments have given to humanity, today anyone who would dare to inject children with virulent pathogens, as Jenner did in the course of his research, would be in trouble.

How much of today's medicine is based on unethical experiments like Jenner's? How much potentially beneficial research is not being carried out because of bioethical restrictions? More important, are the kinds of research that raise bioethical questions necessarily the kinds most capable of helping humanity?

Maybe health research should continue, but with much less emphasis on drug research and much more on other aspects of promoting health. Education for prevention is much less profitable than drugs, perhaps not profitable at all.

Drug research to provide an ounce of cure is more profitable than education for pounds of prevention. So finding funding for education for prevention is more difficult than finding funding for research, but we should still try.

This paper will outline methods for saving the world, including education for disease prevention, the kitchen garden scheme, and individual health autonomy. Among the methods I will discuss under the heading of individual health autonomy are martial arts falling techniques for preventing injury, yoga for improving breathing and the Bates Method for improving eyesight. Although I personally use such methods, my purpose in discussing them is not to recommend them to the reader. I do not give medical advice. I discuss these methods to raise methodological difficulties in testing them, given current methodology in medical research and current philosophy of science. One glaring limitation of these methods is that much of the evidence for them is often anecdotal. Anecdotal evidence, however, is the mother of science, as I will explain at the very end of this paper.

2. Education for Disease Prevention

In a message included in the *2002 World Health Report*, Gro Harlem Brundtland, Director General of the World Health Organization, listed some of the most important risks to health today: high blood pressure, high cholesterol, obesity, underweight, unsafe sex, tobacco consumption, alcohol consumption, unsafe water, sanitation and hygiene, iron deficiency, and interior smoke caused by the burning of solid fuels.[6]

The first three on the list, high blood pressure, high cholesterol, and obesity, may appear to be relevant only to affluent societies. But these and other risk factors that contribute to heart disease are not only concentrated in affluent societies. In 2003, I photographed a billboard advertisement by the TATA-AIG insurance company in Pune, India, that predicted, “By 2010, India will have the highest number of heart patients in the world.”

Studying the risk factors on Brundtland’s list, we can question how relevant pharmaceuticals are to the world’s most significant health problems. As we will see, a more effective response might be education for prevention.

Although high blood pressure, high cholesterol, and obesity might have some genetic components, people can largely prevent or cure these disorders by changes in exercise, diet, and lifestyle.[7] Pharmaceuticals may be appropriate for dealing with the results of unsafe sex. But unsafe sex cannot be prevented by drugs. In developing countries, anti-retrovirals and other AIDS drugs are only reaching a fraction of the people in need.[8] We should focus our efforts more into preventing unsafe sex than into curing its effects. We can prevent unsafe sex only by behavioral change. Although persuading much of the world to adopt permanent virginity might be difficult, promoting condom use might be a more realistic expectation. Perhaps we should direct some funds currently dedicated to AIDS research to developing strategies that encourage women to pressure their sex partners to use condoms. Research into developing such strategies will not be easy. But developing behavior change strategies might be easier, more economical, and more ethical, than doing more anti-retroviral drug research or paying for the drugs developed as a result of that research. In spite of massive efforts by people and governments all over the world, every time I ask health scientists from developing countries about anti-retroviral drugs, I hear that adequate supplies are not getting to the people who need them and that even when they do, compliance is nearly impossible to guarantee.

Some critics might object that despite the need for lifestyle modifications, in clinical practice, efforts to achieve significant lifestyle changes have had limited success. My position is that changing lifestyle is not something we can have doctors do for us. Usually, people have to make lifestyle changes for themselves.

Israel Salanter, founder of the Jewish Mussar Movement, is often quoted as saying that the hardest thing in the world to change is behavioral habits. But charismatic leaders have achieved some notable exceptions. In the 1970s, Padmashree Shri Anna Hazare took voluntary early retirement from the Indian army and went into Ralegan Siddhi, an impoverished village in Maharashtra, India. He organized the villagers, persuaded them to stop alcohol and meat consumption, tobacco use, and gambling. He influenced them to begin hard work in the fields and persuaded them to rise before sunrise for physical exercise before breakfast.[9] This sort of community-wide reform cannot be accomplished by a physician in a clinic.

Largely, underweight is a problem of agricultural methods, land and resource distribution, eating habits, and knowledge about beneficial foods—not a problem that can be solved by drugs. Later in this paper, I will propose the *Kitchen Garden Model* as a solution to some of these causes of underweight. Even so, one major and pharmaceutically treatable cause of underweight in developing countries is infection by parasites.[10] But instead of—or along with—researching drugs to get rid of parasites, we should research ways to avoid being infected with parasites in the first place.

Tobacco and alcohol addiction are not treatable pharmaceutically at this time. The thought that we might develop a pharmaceutical treatment in the future is plausible, though, because alcoholism might have a genetic component that causes an enzyme problem.[11] Once we understand those genetics, enzyme replacement therapy for alcoholism might be developed on the model of enzyme replacement therapy for Gaucher, another heredity enzyme deficiency disorder.

In the meantime, we must seek social solutions, including legal ones. Admittedly, these solutions are difficult. In 2001, Israel passed legislation outlawing smoking in public places. I happened to be in the District Courthouse in Tel Aviv, and went to the coffee shop to get a sandwich. But I could not stay there because the place was full of barristers in black robes, busily puffing on their cigarettes. Walking out, I saw a police officer and tried to make a complaint. He told me that it was not within his jurisdiction because the law against smoking is a municipal bylaw. In Israel, municipal inspectors enforce municipal bylaws, not police. I would have to go to City Hall and file a complaint with the Municipal Inspector. Who has time to do that? Before the new law, Israeli restaurants had smoking and no-smoking sections. Now establishing a smoking section is illegal, so people smoke everywhere. Currently, in Great Britain anti-smoking legislation is being debated.[12] As of this writing—9 September 2005—legislation has not been enacted. The British and other countries contemplating such legislation should learn from the Israeli experience and do a better job. We must ask whether research into how to draft effective anti-smoking legislation, and into how to enforce it, might not save more health and lives than cancer research at equivalent cost.

Social factors contributing to alcohol abuse are complex in developing countries. In the Palar River Delta in Tamil-Nadu, India, *Dalit* (untouchable) men drink *arrack*, strong spirits, which in Asia are chiefly distilled from fermented fruits, grains, or sugarcane, throughout much of their lives. But in the Palar River delta, *arrack* is distilled from brown sugar and fertilizer. In the local dialect of Tamil, fertilizer is the word for agricultural urea. This form of *arrack* causes neurological disease, blindness, and premature death.[13] While the women maintain the house and the fields, one man told us, "Arrack is our life. We make it. We sell it. We drink it."

The social causes of this problem might be related to caste discrimination, resulting in a lack of employment opportunities and poor self-image. I believe research into the social causes and ways to curb its consumption would be more ethical and save more lives than drug research because in the sense of distributive justice, we could foster more health among the medically deprived populations of the world, for whom expensive drugs are not a realistic option.

3. The Kitchen Garden Model

The organic kitchen garden can constitute an integral part of a holistic scheme that can alleviate some of the health problems on Brundtland's list.[14] The model can provide a basis for thought about directions which ethical health research might take. This model has been ubiquitous among organic gardeners for years. What may be new is the presentation of this information within the context of bioethics.

The suggestions in this section are aimed at health needs among villagers in developing countries. City dwellers, rich and poor, may find them less relevant, although more urban compost piles and organic gardens exist now than most people would suspect. The well-known "Swedish dry toilet," which can produce odorless compost, may be particularly suited for urban areas.

Gardening is not a viable option for vast numbers of urban poor or the homeless. Why they do not have garden land available and why they are in the cities in the first place are significant socio-political questions. I hear from friends in developing countries that the urban homeless are homeless because of development, because agricultural, forestry, or mineral developers forced them from their homelands. But the investigation of the extent to which this is true, and what can be done about it, lies beyond the scope of this paper. In any case, vast numbers of rural people in developing countries do have access to tillable land. This section is aimed at those for whom what land is available may be more of a source of health.

Rural doctors in Maharashtra, Ayurvedic (traditional Indian) and Allopathic (conventional Western style), claim that a major cause of health problems among village people is bad nutrition due to not eating enough vegetables. "When they have a little money," one doctor said, "they waste it on meat instead of nutritious vegetables."[15] Not only should the kitchen garden be a partial solution to underweight, but the iron in green, leafy vegetables should help solve problems of iron deficiency.

Some critics object that you cannot get enough protein unless you eat meat, but this claim is not true. Many million strict vegetarians in India know how to maintain health by a proper balance of vegetables with rice, pulses (legumes), and milk products. Not all people who try to be vegetarian are successful. Many make serious attempts to adopt a vegetarian diet but eventually return to eating meat for reasons not entirely easy to state.

Perhaps a lack of knowledge regarding proper mixing of foods in traditional vegetarian diets, or lack of availability of suitable ingredients contributes to their decision. Perhaps a genetic factor plays a role. Perhaps some people have an enzyme that contributes to optimal utilization of the nutrients in grains, pulses, and vegetables. We could test these hypotheses with genetic studies of Brahmin populations that include many million apparently healthy vegetarians.

The fertilizer for the garden should come from composting human, plant, and animal waste. In surveys which our bioethics centre conducted in the Indian States of Tamil-Nadu (in cooperation with the Delta School of Nursing), and Maharashtra (in cooperation with the Jnana Prabhodini organization), we learned that human and animal waste are deposited freely in fields, roadsides and beaches.[16] Instead of leaving this material lying on the ground to be stepped on by barefoot children, to seep down to pollute underground water supplies, or to be carried by rainstorms to streams and ponds, we could compost it. In doing so, we can solve many problems of unsafe water, sanitation and hygiene in developing countries.

In addition, we could eradicate parasites transmitted through human feces to barefoot children, thereby solving many, if not all, causes of underweight. Admittedly, parasitic infections are treatable with drugs. But although drug companies, who control so much of today's medical research, appear to believe that an ounce of cure is worth more profits than pounds of prevention, we should think of returning to the wise adage that an ounce of prevention is worth a pound of cure.

The composting process is already recommended in posters on display in health clinics in the Maharashtra, which has subsidized some home scale bio-gas plants. Composting is carried out in two stages, the first producing bio-gas and the second producing garden fertilizer.

The first stage is anaerobic compost, from which bio-gas, the methane bi-product of anaerobic rotting of organic matter, can be produced. This process is widely known and successful on an industrial scale and, in some parts of the world, on the community and the family scale, where people use methane gas for cooking. In many rural villages, people cook over open fires in unventilated or under-ventilated rooms, causing the problem of interior smoke from burning solid fuels. By encouraging the use of bio-gas instead of solid fuel, we can largely alleviate, if not entirely eradicate, this problem. Indian villagers claim that *chapatti*, the Indian version of pita bread, tastes better prepared over an open fire. In Maharashtra homes we surveyed, families were using home-produced bio-gas for most cooking, but open fires for *chapatti*, a partial solution. Adults and children are still breathing some smoke. But a partial solution is better than no solution. Breathing less smoke is healthier than breathing more.

In the second stage, material already composted anaerobically, thereby having its potential for producing bio-gas exhausted, is mixed with further organic matter and composted again aerobically, to produce a rich, natural garden fertilizer.

Since the methods for the organic kitchen garden already exist and are simple and cheap, the research challenge is not to develop technology but to devise methods for putting existing and widely known low-tech methods into practice. Part of the solution has to be educating vast numbers of people about the values of the kitchen garden scheme. Part has to be political, finding ways to help people gain access to garden land. Even the lack of fertile land should not be a major obstacle because organic soil-building and compost-fertilized raised beds make good gardens possible in the poorest of soils, provided you have water.

Part of the goal of organic gardening is not to use genetically modified seeds. Many critics have objected to genetic engineering on the grounds of safety with the complaint that genetic modification is "playing God." They believe that fundamental characteristics of nature are the work of God, the creation of immutable species on the first six days of creation. Darwinists and their religious opponents have long debated whether species are immutable. The feeling remains among many that we human beings have no right to tamper with species by means of advanced technology.

Insofar as this objection is based on text in the Hebrew Bible, some may be surprised to learn that leading Orthodox rabbis see no problem with tampering with nature if the goal is the good of humanity. Orthodox Jews tend not to see nature as a divine inviolable given. They see human beings, “created in the image” of God and angels as partners in creation. They have no problem with genetic engineering unless good reason exists to believe that the technology is unsafe. Current biomedical literature provides little evidence that genetically engineered foods are unsafe for human consumption. Because not all the data are yet in, people have a right to know what they are eating. Ethics demands that we inform people about the source of the food they buy.

The major problem with genetically modified seeds is that they are designed for the benefit of large-scale agriculture, an industry, which would prefer you buy food at market instead of growing your own. I doubt that family kitchen gardeners have any need whatever for genetically modified (GM) seeds.

We should inquire to what extent modern high-yield varieties of seeds, which were developed before GM seeds, are of value to the kitchen garden or to the small farmer. Varieties and methods which are appropriate for large farm tracts, using modern equipment and artificial fertilizers and pesticides, may not be appropriate for the small, family farm in a developing country. For the family whose land might be too rocky or hilly for modern agricultural equipment, whose sources of fertilizer are only organic, and who use only organic methods, such as companion planting, for pest control, traditional indigenous varieties of seed might be more appropriate. Now organizations such as the Center for Indian Knowledge Systems have created seed banks for preserving traditional, indigenous varieties and making them available to local farmers.

In any case, the kitchen garden scheme is not intended to supply all of a family’s nutritional needs in all cases. Notwithstanding the issue of meat requirements, I would not expect a kitchen garden to supply adequate rice, and probably not adequate pulses either. The value of a kitchen garden remains even when supplemented with other kinds of food.

4. Individual Health Autonomy

Another approach that might obviate the need for much pharmaceutical research is the promotion of health autonomy. Often autonomy in bioethics literature connotes the right of the patient to have a doctor perform euthanasia, to choose to die instead of receiving further treatment for a terminal illness, or to have a doctor provide the means for the patient to commit suicide. We also invoke autonomy to justify the right of patients to be informed about drugs they are taking, about who will be performing surgery on them, about the probability that they might receive a placebo instead of an effective drug in an investigatory trial, and about adverse results observed in clinical trials in which they are still participating.

In addition to these traditional meanings of medical autonomy, the term also means the ability of an individual to decide independently the rules by which he or she will live life. This need not have to do primarily, or even to any large extent, with patients, people who are patiently and passively being treated by active doctors, even if the doctor sometimes acts according to the autonomous opinion of the patient. Neither am I referring primarily to Asian and other non-Western medicine. Asian and other non-Western medicines, along with alternative and complementary techniques, have a place in saving the world. But patients who use these alternatives can be just as passive as one who uses conventional medicines.

Alternative and complementary medicines have a place in saving the world. But patients who uses alternative or complementary medicine can be just as passive one who uses conventional medicines. In this context, by autonomy, I mean an individual exercising responsibility for his or her health, making independent decisions about diet, exercise, lifestyle, and frame of mind, for the purpose of achieving, preserving, or restoring health.

Not every autonomous health decision will be right or smart. Many people will make ill-informed or risky choices about tobacco, whisky, drugs, television, or unprotected sex. But we can encourage people to learn to make more intelligent decisions.

The word "autonomy," when it is used in a medical context, often has to do only with decision making. The phrase "individual health autonomy," as used in this paper, includes decision making but goes much further. To try to make my meaning clearer I will have to go into some metaphysical philosophical ideas, which are not necessary for understanding most of this paper. The reader who is not interested in metaphysics may therefore skip over the following paragraph.

The concept of individual health autonomy is a broadening of the concept of epistemic autonomy, which I proposed in the course of a discussion of Benedict (Baruch) Spinoza's philosophy.[17] According to Spinoza, each mind is a part of God's "attribute of thought," the infinite mind of God. If my mind is a part of God's mind, then I am not thinking my thoughts, wishing my wishes, desiring my desires. God is doing these things and I, as an epistemic being, disappear. The concept of "epistemic autonomy" expresses the belief that our thoughts are not God's thoughts,[18] that our human minds are so separate and different from God's mind that we can even criticize God's ethics, as did Abraham.[19]

The concept of individual health autonomy goes one step further by affirming our individuality in thoughts and decisions, and in taking active conscious control of our own health. Some examples of such conscious control include relaxing muscles, changing our bodies to do a yoga asana, kicking higher than we could ever kick before, relaxing breathing when an asthma attack is threatening, becoming aware of the different effects of different foods on our digestion and health, and changing the way we breathe. The personal aspects of individual health autonomy, as described here, would not be acceptable to proponents of behaviorist philosophy, interested only in those aspects of our mental lives, which are observable to others. Nor would they be acceptable to those who accept Wittgensteinian denials of the possibility of a "private language." Because something is unacceptable to some of us does not entail that the reality does not exist. Instead of the phrase "individual health autonomy," I might have used such a phrase as "conscious, intentional, individual participation," But I have chosen the first phrase to preserve the connection to epistemic autonomy.

This approach to healthful living is individualistic. What works for me might not work for you. So when we try to devise scientific tests of efficacy and safety, we run into some of the same methodological difficulties seen in investigations of Asian and other non-Western medicine. I will discuss some of these methodological difficulties in Section 5.

In what follows, I will discuss in more detail some of methods by which individuals can exercise health autonomy. Later, I will briefly outline the background, in Western philosophy of science, to the basic methodological obstacles of testing forms of individual health autonomy and some forms of complementary and alternative medicine.

Some of these examples are relevant to poor people in developing countries. Part of individual health autonomy is being aware of the effects of different foods, exercise, lifestyle, and environment on digestion and health: Many people, including health professionals, live and eat in an uncontrolled manner, and then go running to a doctor to recover from troubles caused by their lifestyle choices. Most of the a major health risks in Brundtland's list—high cholesterol, obesity, underweight, unsafe sex, tobacco consumption, alcohol consumption, unsafe water, sanitation and hygiene, and iron deficiency—can be controlled by learning how foods and behaviors affect health, then adjusting choices accordingly.

Body awareness is not a new idea. Maimonides' (Moshe ben Maimon, 1135–1204) included extensive discussion of the concept in some of his more popular medical writings, especially the *Guide to Health* and the *Treatise on Asthma.*[20] His ideas greatly relied on distinguished predecessors such as Hippocrates.

Today methods to increase body awareness include meditation, yoga, and Asian martial arts. In all of these activities, students learn to pay careful attention to how they hold their body, what hurts and why. In martial arts, students learn to be aware of and correct how they walk, run, and fall. In all of these activities, students learn to attend to what they eat and how the timing, content, and frequency of meals affect their performance in practicing meditation, yoga, or martial arts. These habits of attention are also beneficial to ordinary activities of daily life.

Some people might think of martial arts only as training for soldiers and police, as a method of self-defense, or as a pastime for people who are attracted to violence. On the contrary, many gentle people whose goals include maintaining a healthy body and a calm, attentive mind, with no interest at all in combat or self-defense applications, practice martial arts. These disciplines can help avoid violence by teaching the self-confidence needed to counter a potential attacker with a smile and calming words. Some—like the ancient Samurai arts and those based on them—also teach how to neutralize an attack without injuring the attacker. These are ways of increasing health by preventing injury.

Another potential health benefit of martial arts is the prevention of injuries caused by falls, sometimes a serious problem not limited to geriatrics. There is a vast medical literature on predicting and preventing falls. But not all falls are preventable. So knowing how to fall safely is beneficial. But the medical literature seems to contain nothing on how to fall safely. Tai Chi teaches balance and "connection to the ground." Tai Chi and other Chinese martial arts do not directly teach safe falling, but Japanese martial arts do.[21] Literature is mounting on the efficacy of training in martial arts to prevent falls. These studies report many positive and some inconclusive results.[22] Although Judo, Aikido, and other Japanese martial arts practitioners testify about the value of their training to taking accidental falls safely. We should devise protocols for testing the hypothesis scientifically, and develop methods for teaching non-athletic, aging people to fall safely.

Some clinical research findings document the benefits of yoga for asthma.[23] Still, physicians who know that overuse of bronchial-dilator inhalers is associated with risk of death or near death still prescribe them for their patients without suggesting alternatives. True, causality has not been established, but why should we use potentially risky drugs when healthy alternatives are worth trying? Even though other literature about the beneficial effects of yoga on asthma is inconclusive, [24] it would behoove the medical profession to investigate these claims.

The Bates Method of improving eyesight without eyeglasses is a perfect example of exercising individual health autonomy because its application is so individualistic.[25] By the Bates Method I mean only the method outlined in William Horatio Bates' book here referenced. His method is different from the self-massage at acupressure points taught in Chinese medicine, and from eye exercises, sometimes taught commercially under his name. The original Bates Method is vastly different from descriptions published by some irresponsible doctors or other critics who have alleged that Bates made outlandish recommendations. For the purpose of this paper, I mean only the Bates Method as described in original materials authored by Bates. I discuss the method not to recommend it, but as a vehicle to discuss methodological problems inherent in attempting to do efficacy research on non-standard methods.

According to Bates, poor vision is not a disease of the eye but a disease of the intellect, a failure of the faculty of imagination. Bates believed that if you can imagine something, you would be able to see what you imagine. He suggested spending time with eyes closed, imagining that you see clearly. Another aid is to spend time looking at large familiar letters. If you can get used to seeing letters the way they should look, it should be easier to do so with small letters as well. If, when trying to read or to look at other things, your vision becomes blurry, you can help it by closing your eyes, imagining that you see it clearly, and then going back to looking at the thing or page. But the most important factor, as with martial arts, is the need to relax. If, according to Bates, you try too hard to see, then your tension will affect the muscles around your eyes, which will distort your lenses, resulting in your seeing even more poorly. You have to reach a state of mind where you do not care whether you see or not. Then, according to his teaching, you will see clearly. The active participation of the individual subject's imagination is what makes Bates' method so hard to test in clinical trials of the generally accepted sort.

For some people, including this author, the Bates Method has helped tremendously, although I must admit that it has not given me back the visual acuity that I had at the age of twenty, and the method may not be effective for serious eye disease. Others find little or no benefit, and yet others—including some extremely learned and intelligent people—cannot even understand Bates' concept.

Because many people who try the Bates Method experience no improvement in vision does not mean that these exercises would not be efficacious for anyone. We will discuss this and other methodological difficulties including the problem of "anecdotal evidence" in the following sections.

5. Methodological Difficulties in Testing Techniques for Exercising Individual Health Autonomy

Aristotle, in the *Posterior Analytics,* argued that for a statement to be scientific, it must be possible to deduce the statement by valid logical reasoning from a general law, usually of the form "all *A*s are *B*" or "no *A*s are *B*." Although different philosophers of science have emphasized different details, this basic form has served as the model for scientific reasoning through the history of philosophy, from the work of David Hume and Immanuel Kant up to the present day.[26]

A major refinement of the model for scientific reasoning, alluded to by Aristotle, but refined by philosophers from the eighteenth century onward, is the element of probability. Modern research results are reported in terms of the probability of obtaining a given result within statistical limits of confidence instead of all-or-nothing conclusions. This refinement is definitely an improvement. To say that something works need not imply that it works for everyone in every situation. It is enough that it should work for a statistically significant segment of the population. But Asian, alternative, and complementary medicine have brought attention to the need to distance ourselves from absolute generalizations. Because something appears to have worked for one person, and appears not to have worked for the rest of the population, need not cause us to deny that it worked for that one person.

When reporting the results of medical research, we can state that yoga appeared to cure Bill of asthma, or that Bates Method appeared to cure Kathy of myopia. But you cannot state that yoga cures asthma or that the Bates Method cures myopia unless you can demonstrate that, using the same methods, similar results are replicable in a significant percentage of a reasonably sized population. You must also be able to demonstrate that these methods are significantly more efficacious than either placebos or other traditional treatments. Otherwise, you have only anecdotal evidence instead of scientific evidence.

This is the epidemiological approach to medicine, dealing with populations—sometimes quite small ones—instead of individuals. We cannot deny the significance of this approach to the development of modern medicine. Still, it has its limitations. No one method has a monopoly on truth. I exaggerate only slightly when I say that according to this approach, you cannot say that something works for you unless you can prove that it works for almost everybody. In addition to the epidemiological approach, health research needs a more individualistic approach, a better understanding of individual cases.

The kitchen garden model is testable under the epidemiological approach because we can design population-wide studies. We have seen that researchers have already studied individual components of the process. All that remains for us to be done is to test it as an integrated system. In the case of other examples that I have discussed, designing traditional efficacy research is difficult.

A. Yoga and Research Methodology

A highly informative review of breathing retraining methods for asthma noted the positive conclusions of one yoga study, but classified it together with other yoga studies, giving the impression that the question is whether yoga in general is effective in relieving asthma symptoms, instead of whether specified kinds of yoga taught by specified teachers to specified pupils are effective. The authors then classified yoga together with other methods of breathing retraining, giving the impression that the more important question is whether breathing retraining is generally effective against asthma. This blurs the specific question whether specific methods of breathing retraining, such as yoga, are more effective than other methods.

Yoga is not always breathing retraining anyway, but is often training in posture and mental attitude, from which changes in breathing are to result without consciously intending them. Yoga instructors often teach a form of breathing exercise called *pranyama* together with yoga. F. S. Ram, E. A. Holloway, and P. W. Jones concluded, "There appears to be little reliable evidence that breathing retraining is of benefit to patients with asthma."[27] Although the authors probably did not intend to encourage doctors to ignore yoga in favor of prescribing inhalers, doctors reading this report could come to this conclusion.

I believe that we must free ourselves of the dogma that only what is easy to test can be scientific. A study of the efficacy of yoga to treat asthma might consider differences among different schools of yoga, among different teachers, and among different individual pupils. There is also a mental aspect to yoga: focusing on designated parts of one's body, directing one's energy to those parts.

I recently told a doctor about my consternation over physicians, knowing the dangers of medicinal inhalers, still prescribing them to their patients without suggesting that they try yoga. She replied that she was not going to send her patients to chase after courses in yoga and like things unless she was convinced that these methods were proven effective. She is right. A patient who needs a doctor to prescribe yoga or Tai Chi or Aikido or meditation is not necessarily the type of person who would succeed in improving their health through these kinds of methods. One who finds these things independently, by one's initiative, and with the perseverance to go through many failures and successes without giving up, just might have the best chance of success. This is not necessarily because of psychosomatic aspects of asthma, but because active methods of seeking health are much harder and require much more individual investment than taking a drug. This individual investment is sometimes painful.

Obviously, these methods are not appropriate for infants. They will also be extremely difficult, perhaps unreachable for people who are too sick or weak to change their way of life. For those who are determined and tenacious enough to take hold of their bodies and those willing to take responsibility for their health, these methods are worth trying.

B. Methodological Difficulties with Bates Method for Improving Eyesight

A well-controlled clinical trial of the Bates Method would be difficult to design because of the imagining factor involved. The subject must relax completely, imagine perfect blackness, and imagine seeing clearly. How can we determine whether the patient attempted and succeeded at these tasks? How can we tell whether someone has relaxed or imagined something, or just thinks that he or she has imagined it? If we steadfastly believe that Bates Method works for all people who do the exercises of imagination as Bates requests, then we could say of those people for whom the method does not work that they did not carry out doctor's orders. But this dogmatic conclusion would be tantamount to presuming the efficacy of the method before the test.

C. Placebo Trials

Designing placebo trials for testing the efficacy of yoga to alleviate asthma and the Bates Method to improve vision without glasses would be a difficult challenge. I cannot conceive of how we could put someone in a situation where he thinks he has imagined something while we (or someone who knows the code) know that he has not. In the case of yoga, ethical questions immediately come to mind. Is telling subjects that they have engaged in yoga (or any other named method), when they have not, ethical? Is it ethical (and safe) to ask people who are dependent on medication to discontinue its use for the duration of a study?

Difficulties also arise in the randomization and double blind aspects of the design. Subjects already accustomed to using inhalers would know their group assignment if investigators were to ask them to discontinue use of the inhaler and to begin some exercise regimen instead. Having had twenty years experience with asthma inhalers before using alternative methods freed me, I know from experience that an inhaler with a placebo would not fool people already familiar with the sensation caused by the authentic drug.

Similarly, random assignment to groups raises ethical, logical, and safety issues: We would not want to deprive asthma sufferers of needed medication before we were sure that the alternative treatment would be effective, yet we cannot test for efficacy without depriving some sufferers of their medication. For some sufferers, their asthma is exercise-induced. In that case, the proposed treatment could exacerbate the condition. Still, with carefully planned safeguards and backup plans, we could compare the effects of yoga on the incidence of asthma attacks to the results of using inhalers or against no treatment at all. We need new thinking in the philosophy of the health sciences and new, imaginative ways to test methods of individual health autonomy.

6. Conclusion: Anecdote, the Mother of Science

My personal experience, or personal experiences related to me by others, forms the basis for much of this paper. Such reports of personal experiences are often termed "anecdotal." The scientific community does not consider anecdotal evidence to be "good science." The World Medical Association Declaration of Helsinki states, "Medical research involving human subjects must conform to generally accepted scientific principles."[28] Judged by this standard, anecdotal reports, which do not conform to what is generally accepted as evidence in medical science, do not appear to be acceptable in medical research.

In contrast, I argue that anecdotal evidence can constitute a legitimate basis for scientific discovery. The etymology of anecdote comes from the Greek, *anekdota*, meaning unpublished items. But the word's meaning has evolved over time. The *Oxford English Dictionary* defines anecdote as meaning "a short or amusing or interesting story about a real incident or person," or "an account regarded as unreliable because based on personal accounts rather than facts or research." For the purpose of this essay, I have taken anecdotal evidence to mean evidence based on observations by one or a few people, which are not necessarily explainable in terms of current generally acceptable scientific principles. I believe that this connotation is adequate to cover most uses of the term by health scientists today.

Three well-known scientific events, which are examples of anecdotal evidence in the sense defined, are:

(1) Albert Abraham Michelson and Edward Williams Morley's investigation of the effect of Earth's rotation on the speed of light, raising questions not answered until Einstein's Theory of Relativity;

(2) Jenner's observation that milkmaids who developed skin rashes on their hands did not fall ill with smallpox, which lead to his discovery of the smallpox vaccine; and

(3) Tycho Brahe's sightings of Mars, which led to Johann Kepler's laws of planetary motion.

I shall comment only on the last of these three examples. Ancient Greek and Hellenistic astronomy assumed that the planets were located in transparent spheres, which moved in perfect circles, and that the speed of a planet was uniform at all stages of its orbit. These apparently elegant perfections of motion lay at the base of the Aristotelian belief that the planets are conscious beings, which contemplate angelic, bodiless intellects and strive to imitate them as they move. For close to two thousand years, people thought this doctrine represented "generally accepted scientific principles."

Thomas S. Kuhn, in *Copernican Revolution*, says of Brahe:

> [He was] the greatest of naked eye observers . . . began the practice of making regular observations of planets as they moved through the heavens rather than observing them only when in some particularly favourable configuration.[29]

Johannes Kepler took especial interest in Brahe's work, especially the sightings which Brahe had recorded of Mars over a twenty years period, "Kepler's scrupulous attempt to fit his orbits to objective data is often cited as an early example of the scientific method at its best."[30] But Kepler, try as he may, was unable to fit Brahe's sightings to the "generally acceptable scientific principle" that each planet moves in a perfect circle at a uniform speed at all times. Kepler finally solved the problem with the first two of his three most famous laws:

> Law 1: The planet[s] move[s] in simple elliptical paths, and the sun occupies one of the two foci of each elliptical orbit.

> Law 2: [T]he orbital speed of each planet varies in such a way that a line joining the planet to the sun sweeps through equal areas of the ellipse in equal intervals of time.[31]

The second law entails that a planet moves faster as it is nearer the sun. This principle violated the old principle of the uniformity of motion.

Brahe's sightings were, strictly speaking, anecdotal; they were observations made by a single person, which were not in line with accepted scientific principles. If Kepler had followed the reasoning held by many of today's medical doctors, he would have rejected Brahe's sightings and stuck blithely with accepted science. Instead, he was genius enough to realize that anecdotal observations should lead us to rethink accepted science.

Today's space travel might have been impossible without Kepler's laws. As Wernher von Braun wrote:

> The exact data which . . . [Kepler] assembled on planetary orbits are part of the essential scientific equipment of astronauts. We use them to calculate the flight paths along which our spacecraft travel in interplanetary space to reach, for instance, a specific landing site on the moon or one of our neighbouring planets. Even the exploratory findings of our spaceships, armed with the latest technical equipment, have not succeeded in reversing the laws established by Kepler, but in endorsing them.[32]

It is perhaps unfair to associate the dogmatic rejection of anecdotal evidence with Western medicine in general. Some doctors do recognize that anecdote "is often a more powerful persuader than scientific publication in changing clinical practice."[33] Yet not enough doctors and researchers recognize that the anecdotal evidence of the vast numbers of people who prefer alternative, complementary, and Asian medicine to the Western variety, or who prefer to try to look after their own health autonomously, calls for a new Keplerian revolution in the search for new methods to test and study this evidence.

ACKNOWLEDGMENT

I wish to thank Shimon Glick, for tough criticism of an earlier draft. I have incorporated many of his phrases into the present version. This is not to say that the present version will be immune to his criticism, or that he will agree with even a little of what I say. I have not been able to convince him of my opinions in years of debate. Nevertheless, I am trying.

NOTES

1. Frank J. Leavitt, "Is Any Medical Research Population Nonvulnerable?" Cambridge Quarterly of Healthcare Ethics, 15 (2006), pp. 81–88.

2. K. J. Parkins, C. F. Poets, M. O'Brien M, V. A. Stebbens, and Southall, "Effect of Exposure to 15 percent Oxygen on Breathing Patterns and Oxygen Saturation in Infants: Interventional Study," *British Medical Journal*, 316 (1998), pp. 887–891; R. S. Glock, and J. R. Goldim, "Informed Consent in Gerontology," *Eubios Journal of Asian and International Bioethics*, 13 (2003), pp. 6–8; and E. J. Cassell, A. C. Leon, and S. G. Kaufman, "Preliminary Evidence of Impaired Thinking in Sick Patients," *Annals of Internal Medicine*, 134 (2001), pp. 1120–1123.

3. John Colquhoun, "Why I Changed My Mind about Water Fluoridation," *Perspectives in Biology and Medicine*, 41:1 (Autumn 1997), pp. 29–44.

4. J. N. Rao, and L. J. Sant Cassi, "Ethics of Undisclosed Payments to Doctors Recruiting Patients in Clinical Trials," *British Medical Journal*, 325 (2002), pp. 36–37.

5. Jon H. Thompson, Jocelyn Grant Downie, and P. Baird, The Olivieri Report: The Complete Text of the Report of the Independent Inquiry Commissioned by the Canadian Association of University Teachers (Toronto, James Lorimer, 2001).

6. World Health Association, *The World Health Report, 2002: Reducing Risks, Promoting Healthy Life* (Geneva, Switzerland: World Health Organization, 2002).

7. Richard A. Dickey and John J. Janick, "Lifestyle Modifications for Prevention and Treatment of Hypertension," *Endocrine Pracitce*, 7:5 (2004), pp. 392–399.

8. A. S. Furber, I. J. Hodgson, A. Desclaux, and D. S. David S Mukasa, "Barriers to Better Care for People with AIDS in Developing Countries," *British Medical Journal*, 329 (2004), pp. 1281–1283.

9. Anna Hazare, *A Veritable Transformation*, trans. B. S. Pendse (Ralegan Siddhi, India: Ralegan Siddhi Pariwar Publications, 1997).

10. L. S. Stephenson, "Helminth Parasites: A Major Factor in Malnutrition," *World Health Forum*, 15 (1994), pp. 169–172.

11. L. Lumeng and D. W. Crabb, "Genetic Aspects and Risk Factors in Alcoholism and Alcoholic Liver Disease," *Gastroenterology*, 107 (1994), pp. 572–578.

12. Claire Laurent. "UK Health Minister under Pressure to Ban Smoking in Public," *British Medical Journal*, 329 (2004), p. 368.

13. A. Tanabe and Frank J. Leavitt, Center for Asian and International Bioethics Team, Interviews with villagers and with the staff of the Delta School of Nursing, Kadalur, Tamil-Nadu, India, 2000, 2001, http://fohs.bgu.ac.il/toplevel/TDetails.asp?Div Type=CNT&DivID=3600 (accessed 18 October 2005).

14. Frank J. Leavitt and Mrinalinee Vanarase. "Making Dirty Water Drinkable: Kitchen Garden Scheme is Alternative" (Letter), *British Medical Journal*, 331 (2005), p. 781.

15. A. Tanabe, S. Kulkarni, and Frank J. Leavitt, Health Survey, Velhe Block, Pune District, Maharashtra, India, 2001 http://fohs.bgu.ac.il/toplevel/TDe tails.asp?Div Type=CNT&DivID=3600 (accessed 18 October 2005).

16. Leavitt and Vanarase, "Making Dirty Water Drinkable."

17. Frank J. Leavitt, "Weeks, Spinoza's God and Epistemic Autonomy," Sophia: A *Journal for Philosophical Theology and Cross-Cultural Philosophy of Religion*, 31 (1992), pp. 111–118.

18. Isaiah, 55:8.

19. Genesis, 18:25.

20. Maimonides, Moses, *The Preservation of Youth: Essays on Health*, trans. (from Arabic) Fi Tadbir Assihha, (New York: Philosophical Library, 1958); in Hebrew: as *Moshe ben Maimon, Ketavim Refu'iyim (Medical Works)*, ed. Suessmann Muntner (Jerusalem: Mossad ha-Rav Kuk, 1957).

21. Laurie L. Lavery and Stephanie A. Studenski, "Tai Chi, Falls, and the Heritage of JAGS," *Journal of the American Geriatrics Society*, 51:12 (2003), pp. 1804–1805; and S. L. Wolf, R. W. Stattin, M. Kutner, et al., "Intense Tai Chi Exercise Training and Fall Occurrences in Older, Transitionally Frail Adults: A Randomized, Controlled Trial," *Journal of the American Geriatrics Society*, 51 (2003), pp. 1693–1701.

22. Frank J. Leavitt, "Can Martial Arts Falling Techniques Prevent Injuries?" Research Letter, *Injury Prevention*, 284 (2003), p. 9.

23. R Nagarathna and H. R. Nagendra, "Yoga for Bronchial Asthma: A Controlled Study," *British Medical Journal*, 291 (1985), pp. 1077–1079.

24. P. K. Vedanthan, L. N. Kasavalu, and K. C. Mutthy, "Clinical Study of Yoga Techniques in University Students with Asthma: A Controlled Study," *Allergy Asthma Proceedings*, 19 (1998), pp. 3–9; and W. O. Spitzer, S. Suissa, P. Ernst, R. I. Horwitz, B. Habbick, D. Cockcroft, J. F. Boivin, M. McNutt, A. S. Buist, and A. S. Rebuck, "The Use of Beta-Agonists and the Risk of Death and Near Death from Asthma," *New England Journal of Medicine*, 326 (1992), pp. 501–506.

25. William Horatio Bates, *The Bates Method for Better Eyesight without Glasses* (London, Grafton, 1979).

26. Frank J. Leavitt, "Aristotle's Theory of Explanation" (in Hebrew), *Ivvun*, 37 (1988), pp. 207–214; and Frank J. Leavitt, "Kant's Schematism and His Philosophy of Geometry," *Studies in History and Philosophy of Science*, 22 (1991), pp. 647–659.

27. F. S. Ram, E. A. Holloway, and P. W. Jones, "Breathing Retraining for Asthma," *Respiratory Medicine*, 97 (2003), pp. 501–507.

28. World Medical Association, "Declaration of Helsinki: Ethical Principles for Medical Research Involving Human Subjects, Edinburgh, Scotland, October 2000," *World Medical Association, Recommendations Guiding Physicians in Biomedical Research Involving Human Subjects* (Ferney-Voltaire Cedex, France: World Medical Association, 2002).

29. Thomas S. Kuhn, *The Copernican Revolution: Planetary Astronomy in the Development of Western Thought* (New York: Vintage Books, 1957/1959), p. 200.

30. Ibid., p. 215.

31. Ibid., p. 212.

32. Wernher von Braun, "Kepler and Space Travel," *Johannes Kepler 1571/1971*, eds. Wernher von Braun and Friedrich Abel (Bonn-Bad Godesberg, Internationes, 1971), pp. 5–7.

33. Judith Green and Nicky Britten, "Qualitative Research and Evidence Based Medicine," *British Medical Journal*, 316 (1998), pp. 1230–1232.

ABOUT THE AUTHORS

RICHARD ASHCROFT, MA (Cantab), PhD, is Reader in Biomedical Ethics at Imperial College of London. He has written extensively on the ethics of medical research and has been a member and a trainer of research ethics committees in the UK and overseas. He is one of the editors of Case Analysis in Clinical Ethics, Eds. A. Ashcroft, A. Lucassen, M. Parker, M. Verkerk, M and G. Widdershoven (Cambridge: Cambridge University Press, 2005).

MARGARET P. BATTIN, MFA, PhD, is Distinguished Professor in the Department of Philosophy at the University of Utah, and Adjunct Professor in the Division of Medical Ethics, Department of Internal Medicine at the University of Utah School of Medicine. Her current areas of interest and research include suicide, physician-assisted suicide and euthanasia, ethics and infectious disease, and reproductive rights. Her most recent book is *Ending Life: Ethics and the Way We Die*, published by Oxford in 2005.

SOLOMON R. BENATAR, MBChB, FFA (SA), FRCP, is Professor of Medicine and Founding Director of the Bioethics Centre at the University of Cape Town. He was Chairman of the University of Cape Town's Department of Internal Medicine and Chief Physician at Groote Schuur Hospital in Cape Town during 1980–1999, and President of the International Association of Bioethics during 2001–2003. Currently, he is Visiting Professor in Public Health Sciences and Medicine at the University of Toronto and Director of the University of Toronto and Director of a Fogarty International Center funded program (at the University of Cape Town) on capacity building in International Research Ethics in southern Africa. His academic interests range from respiratory medicine, academic freedom, medical ethics, and the humanities in medicine, to human rights, health-care systems, health economics, and global health. He has published over 230 journal articles and book chapters. He has been elected a Foreign Associate Member of the United States National Academy of Science Institute of Medicine, an Honorary Foreign Member of the American Academy of Arts and Sciences, and Fellow of the Royal Society of South Africa.

JEFFREY R. BOTKIN, MD, MPH, is Professor of Pediatrics at the University of Utah, an Adjunct Professor of Internal Medicine in the Division of Medical Ethics, and an Adjunct Professor of Human Genetics. He is the Associate Vice President for Research Integrity at the University of Utah. His research focuses on the ethical, legal, and social implications of genetic technology with emphasis on research ethics, genetic testing for cancer susceptibility, newborn screening, and prenatal diagnosis.

ALEXANDER MORGAN CAPRON, LLB, is the first Director of Ethics, Trade, Human Rights, and Health Law at the World Health Organization, Geneva, Switzerland. He previously taught law, medicine, and ethics at Georgetown University, University of Pennsylvania, Yale University, and most recently at the University of Southern California where he was University Professor, Henry W. Bruce Professor of Equity, Professor of Law and Medicine, and Co-Director of the Pacific Center for Health Policy and Ethics. He is President of the International Association of Bioethics (IAB), and has been a Public Member of the Board of Commissioners of the Joint Commission on Accreditation of Healthcare Organizations (JCAHO), since 1994. He has written or edited eight books, including *Law, Science and Medicine* and the *Treatise on Health Care Law*.

LEONARDO D. DE CASTRO, AB, MA, PhD, is Professor of Philosophy at the University of the Philippines. He is Vice Chair of the UNESCO International Bioethics Committee, Secretary of the International Association of Bioethics, Vice Chair of the Forum for Ethics Review Committees in the Asia-Pacific Region, and Vice President of the Asian Bioethics Association. He has been the recipient of a National Book Award and the Takashi Fujii Prize given by the International Federation of Social Science Organizations. A member of the National Ethics Committee of the Philippines since 1996, Dr. de Castro also sits on the Philippine Organ Donation Advisory Board and chairs the Medical City Research Ethics Committee.

LESLIE PICKERING FRANCIS, PhD, JD, is Professor and Chair, Department of Philosophy, and Alfred C. Emery Professor of Law at the University of Utah. She is most recently an author of "How Infectious Diseases Got Left Out and What this Omission Might have Meant for Bioethics," *Bioethics* (2005), and coauthor of "Justice through Trust: Disability and the 'Outlier Problem' in Social Contract Theory," *Ethics* (2005). She has written widely on issues of justice in health care, on many topics in health law, and on ethical and legal issues involving persons with disabilities.

BART HANSEN, MA, is a research assistant at the Fund for Scientific Research–Flanders, on the Faculty of Medicine at the Center for Biomedical Ethics and Law, Katholieke Universiteit Leuven. He is also a member of the Faculty of Theology at Leuven, and is currently completing a doctoral research project on the ethical and theological issues surrounding human stem cells research.

MATTI HÄYRY, BA, MA, MSc, LicSc, DSc, is Professor of Bioethics and Philosophy of Law at the University of Manchester and Vice President of the International Association of Bioethics during 2005–2007. He studied philosophy at the University of Helsinki and held several research and teaching positions in Finnish universities before moving to the United Kingdom in 2001. Before taking up his present position, he was Professor of Moral Philosophy and Head of Center for Professional Ethics at the University of Central Lancashire. His collaboration in Finland continues—he is Adjunct Professor of Practical Philosophy at the University of Helsinki and he coordinates a project on the Ethical and Social Aspects of Bioinformatics, financed during 2004–2007 by the Academy of Finland. He has authored, co-authored, and co-edited twenty-four books and collective works, including *Critical Studies in Philosophical Medical Ethics* (1990), *Liberal Utilitarianism and Applied Ethics* (1994), *Playing God: Essays on Bioethics* (2001), *Scratching the Surface of Bioethics* (2003), and *Bioethics and Social Reality* (2005); and published about 180 articles on bioethics, social ethics, and general philosophy in academic journals and edited collections. He is currently working on a book tentatively titled *Improving Humanity? Genetic Challenge and the Unbearable Fuzziness of Ethics*.

PETER HERISSONE-KELLY is Lecturer in Professional Ethics in the Centre for Professional Ethics at the University of Central Lancashire, UK. He has a BA in Philosophy from Bolton Institute, and a BPhil in Philosophy from the University of Oxford. He is currently working on a PhD dissertation titled, "Situations, Reasons, and Incentives: Kant on Rational Agency and Moral Motivation," and has published articles in the *Journal of Medical Ethics*, the *Cambridge Quarterly of Healthcare Ethics*, and *Kant-Studien*.

JAY A. JACOBSON, MD, is Professor of Internal Medicine, Chief of the Division of Medical Ethics, and a member of the Division of Infectious Disease at LDS Hospital and the Department of Medicine at the University of Utah, School of Medicine. His current areas of interest and research include clinical and classroom teaching of medical ethics, infectious diseases, subspecialty consultation and patient care, end of life care, and informed consent and medical decision making.

ALASTAIR KENT, BA (Hons.), MA (Cantab), MPhil, is the Director of the Genetic Interest Group (GIG), a UK alliance of charities and support groups for people affected by genetic disorders. GIG's mission is to promote the development of the scientific understanding of genetics and the part that genetic factors play in health and disease, and to see the speedy transfer of this new knowledge into improved services and support for the treatment of currently incurable conditions. Prior to joining GIG, Alastair worked for several voluntary organizations on issues concerning policy, service development and disabled people.

FRANK J. LEAVITT, PhD, (also known as Yeruham) teaches philosophy and ethics at the Faculty of Health Sciences, Ben Gurion University of the Negev, Beer Sheva, Israel. At other times, he has been a bicycle mechanic, dairy goat man, glazier, desert agricultural worker, organic gardener, yeshiva student, and Israeli settler. He is interested in martial arts and other non-medical methods of promoting health and spirituality. He and his wife, June, have five children and three grandchildren, and hope to have many more.

FLORENCIA LUNA, MA, PhD, is Adjunct Researcher at the National Scientific and Technological Research Council (CONICET), Argentina. She is directing the area of bioethics at Facultad Latinamericana de Ciencias Sociales (FLACSO) and is co-director with Ruth Macklin of a United States National Institute of Health research training grant. She was President of the International Association of Bioethics during 2003–2005, and she is Temporary Advisor of the World Health Organization (WHO). She was a member of the Steering Committee of the Council for International Organizations of Medical Sciences (CIOMS) that worked on the International Ethical Guidelines for Biomedical Research Involving Human Subjects. She has been the editor of *Perspectivas Bioéticas*, the first Argentinian journal wholly devoted to bioethics, since 1996. She has also published articles in national and international journals. She is author of *Ensayos de Bioética: Reflexiones desde el Sur* (2001) and co-author of the books *Decisiones de Vida y Muerte* (1995) and *Bioetica* (1998). Presently she is working on issues related with research in developing countries and ethics, genetics, and international codes of ethics.

DARRYL R. J. MACER holds a BSc (Hons) in Biochemistry from Lincoln College, University of Canterbury (1983), and a PhD in Biochemistry at the MRC Laboratory of Molecular Biology, and Trinity College, University of Cambridge, U.K. (1987). He has been the Director of Eubios Ethics Institute, New Zealand, since 1990. He is Associate Professor, teaching bioethics, at the Institute of Biological Sciences, University of Tsukuba, Tsukuba Science City, Japan, and Visiting Professor at United Nations University since 2002. He is the Founding Director of the International Union of Biological Sciences (IUBS) Bioethics Program, since 1997; Member of the UNESCO Bioethics Committee during 1993–1998; Member, Human Genome Organisation (HUGO) Ethics Committee, UK, since 1995; Board Member of the International Association of Bioethics since 1999; Secretary of the Asian Bioethics Association; and the Editor of the *Eubios Journal of Asian and International Bioethics* since 1990. He has authored eight books and edited nine books in English, edited seven books in Japanese, and authored over 150 academic papers. His current projects include the Human Behaviorome Project at the Eubios Ethics Institute in Japan, and he is developing bioethics education materials. His recent books include *Bioethics is Love of Life: An Alternative Text-*

book (1998), *Bioethics for the People by the People* (1994), and *Bioethics in High Schools in Australia, Japan and New Zealand.*

LEONARD M. MARTIN, MA, PhD, CSsR, did his master's work in moral theology and his doctoral work in medical ethics. He was ordained a Roman Catholic priest in 1977, and was a member of the Redemptorist Order of Dublin. He was also a Member of the National Commission for Ethics in Research (CONEP) during 1997–2004, and became Professor of Bioethics at the Public University of Ceará (Fortaleza) in the North-East part of Brazil in 2001. In recent years, Leonard became well-known for his writings in bioethics, and he had become a specialist on the problem of AIDS. He traveled extensively giving lectures in both these fields. He died after a brief illness in 2004.

MICHAEL PARKER, BEd, MA, PhD, is Professor of Bioethics and Director of the Ethox Center at the University of Oxford. He is also Honorary Clinical Ethicist at the Oxford Radcliffe Hospitals Trust, providing ethics support and education in the clinical setting, especially in genetics. His main research interests include the use of multidisciplinary research to investigate the ethical and social implications of biotechnology, ethical issues in genetic epidemiology on malaria in developing countries, and ethical issues arising in the clinical use of genetics. He is also doing research into ethical issues arising in research on patient records without patients' consent. He is a principal investigator on the Oxford Genetics Knowledge Park leading a multidisciplinary team investigating the ethical, legal, psychological, economic, and social aspects of genetics research and practice. He is also a partner in the MalariaGEN Project, funded jointly by the Foundation for the National Institute of Health and Wellcome Trust. He is co-editor with Julian Savulescu of "Ethics in Practice," *British Medical Journal*, and a member of the editorial team of the *Journal of Medical Ethics*. He is the author of several peer-reviewed papers and books on bioethics, and is one of the editors of Case Analysis in Clinical Ethics, Eds. A. Ashcroft, A. Lucassen, M. Parker, M. Verkerk, M and G. Widdershoven (Cambridge: Cambridge University Press, 2005).

MARIA DO CÉU PATRÃO NEVES is Professor of Philosophy, specializing in Ethics and Philosophical Anthropology, at the University of the Azores in Portugal. She is a Member of the National Ethics Committee and of the Board of Directors of the International Association of Bioethics. Working in the field of Bioethics since the late 1980s, she did postdoctoral studies during 1992–1993, and a sabbatical year during 1999–2000, at the Kennedy Institute of Ethics in Washington, D.C. She teaches regularly at the Bioethics Master's Degree Program of the Schools of Medicine and the Bioethics Institute at Catholic University of Lisbon and Oporto. She is Head of the Center for Bioethics Studies at the University of Azores in Portugal, and member of a Hospital Ethics Committee and of the European Ethics Network. She is a member of the Editorial Boards of *Medicine, Health Care, and Philosophy* (Nijmegen, Holland) and *Ethical Perspectives* (Leuven, Belgium). She has led the organization committees of many bioethics meetings, some of them international. She has authored several books including *Comissões de Ética: Das Bases Teóricas à Actividade Quotidiana* (1996, 2nd ed., 2002); *Para uma Ética da Enfermagem: Desafios* (2004) and about eighty papers on philosophy and on bioethics.

LEO PESSINI, PhD, is Professor of Bioethics at the St. Camillus University Center, São Paulo, Brazil, specializing in Moral Theology. He was a Founder Member of the National Commission of Ethics in Research (CONEP) of Brazil, and he is the Vice-President of the Brazilian Bioethics Society.

PAUL SCHOTSMANS is Professor of Medical Ethics at the Faculty of Medicine of the Catholic University of Leuven, since 1984. He is a Priest of the Archdiocese Mechelen-Brussels and Doctor in Theology from the Katholieke Universiteit Leuven, Belgium. Since 1986, he is Director of the Center for Biomedical Ethics and Law at the Department of Public Health of the same university. He served as President of the European Association of Centers of Medical Ethics and is elected Board Member of the International Association of Bioethics. Since 1996, he has been a full member of the Belgian Advisory Committee on Bioethics.

CHARLES B. SMITH, MD, is Professor Emeritus of Medicine at the University of Washington, School of Medicine. He previously served as Professor and Associate Dean. He is currently Professor Emeritus of Medicine at the University of Utah, School of Medicine, where he has held positions as Associate Chairman, Professor of Medicine, and Chief of the Division of Infectious Diseases. His current research interests are in the field of medical ethics.

TUIJA TAKALA, MSc (soc), LicSc (soc), PhD is Lecturer in Bioethics and Moral Philosophy at the University of Manchester, United Kingdom, and Adjunct Professor in Practical Philosophy at the University of Helsinki, Finland. Dr. Takala's research interests lie in the philosophical analysis of bioethical arguments and concepts, and she has a special interest in issues related to genetic information, consent, autonomy, and paternalism. Her publications include contributions to *Bioethics*; *Cambridge Quarterly of Healthcare Ethics*; *Journal of Medical Ethics*; *Journal of Medicine and Philosophy*; *Medicine, Healthcare and Philosophy*; *Perspectivas Bioéticas*; *Theoretical Medicine and Bioethics*; *Western Journal of Medicine*, and the book *Scratching the Surface of Bioethics* (ed. with Matti Häyry) (New York and Amsterdam: Rodopi, 2003). Takala is the chair of the editorial board of the *Cambridge Quarterly of Healthcare Ethics*, co-editor of the *Values in Bioethics* special book series

(Rodopi), and member of the editorial board of the journal *Theoretical Medicine and Bioethics*.

ROSEMARIE TONG received her PhD in Philosophy from Temple University in 1978. She is currently the Distinguished Professor of Health Care Ethics in the Department of Philosophy and the Director of the Center for Professional and Applied Ethics at the University of North Carolina at Charlotte. She is the Past Coordinator of the International Network of Feminist Approaches to Bioethics and currently serves as an Executive Board Member of the International Association of Bioethics. She is the Chair of the American Philosophical Association's Committee on the Status of Women, a frequent panellist and judge for the National Institute of Health, and a winner of the 1986 Council for Advancement and Support of Education (CASE) National Professor of the Year Award. She has authored or co-edited thirteen books, including *Controlling Our Reproductive Destiny: A Technological and Philosophical Perspective* (1994), *Feminist Approaches to Bioethics* (1996), *Feminist Thought: A More Comprehensive Introduction* (1998), *Globalizing Feminist Bioethics: Crosscultural Perspectives*, with Aida Santos and Gwen Anderson (2001), and *Linking Visions: Feminist Bioethics, Human Rights, and the Developing World* (with Anne Donchin and Sue Dodds, 2004). She has published over 100 articles on topics related to reproductive and genetic technology, biomedical research, feminist bioethics, and global bioethics. She is a frequent speaker at national and international conferences, universities, and medical schools.

ANGELA AMONDI WASUNNA is the Associate for International Programs at The Hastings Center, New York. She is a lawyer by training and an Advocate of the High Court of Kenya. Angela received a Bachelor of Laws degree from the University of Nairobi Kenya in 1996, and two Master of Laws degrees (with bioethics specializations) from McGill University and Harvard Law School in 2000. Angela's research and writing interests include international health law, health and human rights issues, reproductive law and policy, financing of health care in developing countries, ethical issues raised by international research and HIV/AIDS. Angela is a Board Member of the International Association of Bioethics, a member of the International Bar Association, the Pan African Bioethics Initiative, and the Law Society of Kenya. She has just completed a book, *Cost, Choice and Equity: Medicine and the Market* (with Daniel Callahan, PhD, forthcoming from Johns Hopkins University Press, 2005).

ANDREW WILKIE, BM, BCh, MA, DM, FRCP, FmedSci, is the Nuffield Professor of Pathology at the Weatherall Institute of Molecular Medicine, Oxford, an Honorary Consultant in Clinical Genetics, and Chairman of the Oxford Genetics Knowledge Park. His research focuses on genetic causes of skull and limb malformations.

BORIS G. YUDIN was born in 1943 in Moscow, Russia. He is a Corresponding Member of the Russian Academy of Sciences and Vice-Chairman of the Russian National Committee on Bioethics, Russian Academy of Sciences. He is the Editor-in-Chief of the scientific and popular journal *Chelovek* (*The Human Being*), the leading Russian periodical dealing with bioethical issues, and he is the Russian representative on the Council of Europe Steering Committee on Bioethics. He is the editor, author, and co-author of several chapters in *Introduction to Bioethics*—the first textbook on bioethics published in Russian, and he has been a participant in several World Congresses on bioethics. Topics of his most recent publications include ethical problems and regulations in experiments with human beings and in human genome research, human cloning—ethical issues and public perception, genetic and social construction of human life, technological advances and human nature, and different value orientations in bioethics.

ELMA LOURDES CAMPOS PAVONE ZOBOLI, Nurse, Master of Bioethics, Doctor of Public Health Sciences, is Professor at University of São Paulo Nursing School in Brail. She is a Member of the Board of Directors of the International Association of Bioethics and Secretary of the Brazilian Society of Bioethics. During 1997–2003, she was a Member of the Brazilian Health Ministry National Commission for Research Ethics. She is author and co-editor of books, journal articles, and chapters in books on ethics and healthcare management, bioethics and public health, biomedical research ethics, and promotion of health. She was the recipient of a Brazilian Health Ministry award for her doctoral thesis titles "Bioethics and Primary Care: A Descriptive Ethics Study with Nurses and Doctors from the Family Health Program" (University of São Paulo, Brazil, 2003).

INDEX

VIBS

The **Value Inquiry Book Series** is co-sponsored by:

Adler School of Professional Psychology
American Indian Philosophy Association
American Maritain Association
American Society for Value Inquiry
Association for Process Philosophy of Education
Canadian Society for Philosophical Practice
Center for Bioethics, University of Turku
Center for Professional and Applied Ethics, University of North Carolina at Charlotte
Central European Pragmatist Forum
Centre for Applied Ethics, Hong Kong Baptist University
Centre for Cultural Research, Aarhus University
Centre for Professional Ethics, University of Central Lancashire
Centre for the Study of Philosophy and Religion, University College of Cape Breton
Centro de Estudos em Filosofia Americana, Brazil
College of Education and Allied Professions, Bowling Green State University
College of Liberal Arts, Rochester Institute of Technology
Concerned Philosophers for Peace
Conference of Philosophical Societies
Department of Moral and Social Philosophy, University of Helsinki
Gannon University
Gilson Society
Haitian Studies Association
Ikeda University
Institute of Philosophy of the High Council of Scientific Research, Spain
International Academy of Philosophy of the Principality of Liechtenstein
International Association of Bioethics
International Center for the Arts, Humanities, and Value Inquiry
International Society for Universal Dialogue
Natural Law Society
Philosophical Society of Finland
Philosophy Born of Struggle Association
Philosophy Seminar, University of Mainz
Pragmatism Archive at The Oklahoma State University
R.S. Hartman Institute for Formal and Applied Axiology
Research Institute, Lakeridge Health Corporation
Russian Philosophical Society
Society for Existential Analysis
Society for Iberian and Latin-American Thought
Society for the Philosophic Study of Genocide and the Holocaust
Unit for Research in Cognitive Neuroscience, Autonomous University of Barcelona
Yves R. Simon Institute

Titles Published

1. Noel Balzer, *The Human Being as a Logical Thinker*

2. Archie J. Bahm, *Axiology: The Science of Values*

3. H. P. P. (Hennie) Lötter, *Justice for an Unjust Society*

4. H. G. Callaway, *Context for Meaning and Analysis: A Critical Study in the Philosophy of Language*

5. Benjamin S. Llamzon, *A Humane Case for Moral Intuition*

6. James R. Watson, *Between Auschwitz and Tradition: Postmodern Reflections on the Task of Thinking.* A volume in **Holocaust and Genocide Studies**

7. Robert S. Hartman, *Freedom to Live: The Robert Hartman Story*, Edited by Arthur R. Ellis. A volume in **Hartman Institute Axiology Studies**

8. Archie J. Bahm, *Ethics: The Science of Oughtness*

9. George David Miller, *An Idiosyncratic Ethics; Or, the Lauramachean Ethics*

10. Joseph P. DeMarco, *A Coherence Theory in Ethics*

11. Frank G. Forrest, *Valuemetrics*$^{\aleph}$: *The Science of Personal and Professional Ethics.* A volume in **Hartman Institute Axiology Studies**

12. William Gerber, *The Meaning of Life: Insights of the World's Great Thinkers*

13. Richard T. Hull, Editor, *A Quarter Century of Value Inquiry: Presidential Addresses of the American Society for Value Inquiry*. A volume in **Histories and Addresses of Philosophical Societies**

14. William Gerber, *Nuggets of Wisdom from Great Jewish Thinkers: From Biblical Times to the Present*

15. Sidney Axinn, *The Logic of Hope: Extensions of Kant's View of Religion*

16. Messay Kebede, *Meaning and Development*

17. Amihud Gilead, *The Platonic Odyssey: A Philosophical-Literary Inquiry into the Phaedo*

18. Necip Fikri Alican, *Mill's Principle of Utility: A Defense of John Stuart Mill's Notorious Proof.* A volume in **Universal Justice**

19. Michael H. Mitias, Editor, *Philosophy and Architecture.*

20. Roger T. Simonds, *Rational Individualism: The Perennial Philosophy of Legal Interpretation.* A volume in **Natural Law Studies**

21. William Pencak, The Conflict of Law and Justice in the Icelandic Sagas

22. Samuel M. Natale and Brian M. Rothschild, Editors, *Values, Work, Education: The Meanings of Work*

23. N. Georgopoulos and Michael Heim, Editors, *Being Human in the Ultimate: Studies in the Thought of John M. Anderson*

24. Robert Wesson and Patricia A. Williams, Editors, *Evolution and Human* Values

25. Wim J. van der Steen, *Facts, Values, and Methodology: A New Approach to Ethics*

26. Avi Sagi and Daniel Statman, *Religion and Morality*

27. Albert William Levi, *The High Road of Humanity: The Seven Ethical Ages of Western Man*, Edited by Donald Phillip Verene and Molly Black Verene

28. Samuel M. Natale and Brian M. Rothschild, Editors, *Work Values: Education, Organization, and Religious Concerns*

29. Laurence F. Bove and Laura Duhan Kaplan, Editors, *From the Eye of the Storm: Regional Conflicts and the Philosophy of Peace.* A volume in **Philosophy of Peace**

30. Robin Attfield, *Value, Obligation, and Meta-Ethics*

31. William Gerber, *The Deepest Questions You Can Ask About God: As Answered by the World's Great Thinkers*

32. Daniel Statman, *Moral Dilemmas*

33. Rem B. Edwards, Editor, *Formal Axiology and Its Critics*. A volume in **Hartman Institute Axiology Studies**

34. George David Miller and Conrad P. Pritscher, *On Education and Values: In Praise of Pariahs and Nomads*. A volume in **Philosophy of Education**

35. Paul S. Penner, *Altruistic Behavior: An Inquiry into Motivation*

36. Corbin Fowler, *Morality for Moderns*

37. Giambattista Vico, *The Art of Rhetoric* (*Institutiones Oratoriae*, 1711–1741), from the definitive Latin text and notes, Italian commentary and introduction byGiuliano Crifò.Translated and Edited by Giorgio A. Pinton and Arthur W. Shippee. A volume in **Values in Italian Philosophy**

38. W. H. Werkmeister, *Martin Heidegger on the Way*. Edited by Richard T. Hull. A volume in **Werkmeister Studies**

39. Phillip Stambovsky, *Myth and the Limits of Reason*

40. Samantha Brennan, Tracy Isaacs, and Michael Milde, Editors, *A Question of Values: New Canadian Perspectives in Ethics and Political Philosophy*

41. Peter A. Redpath, *Cartesian Nightmare: An Introduction to Transcendental Sophistry*. A volume in **Studies in the History of Western Philosophy**

42. Clark Butler, *History as the Story of Freedom: Philosophy in InterculturalContext*, with responses by sixteen scholars

43. Dennis Rohatyn, *Philosophy History Sophistry*

44. Leon Shaskolsky Sheleff, *Social Cohesion and Legal Coercion: A Critique of Weber, Durkheim, and Marx*. Afterword by Virginia Black

45. Alan Soble, Editor, *Sex, Love, and Friendship: Studies of the Society for the Philosophy of Sex and Love, 1977–1992.* A volume in **Histories and Addresses of Philosophical Societies**

46. Peter A. Redpath, *Wisdom's Odyssey: From Philosophy to Transcendental Sophistry*. A volume in **Studies in the History of Western Philosophy**

47. Albert A. Anderson, *Universal Justice: A Dialectical Approach*. A volume in **Universal Justice**

48. Pio Colonnello, *The Philosophy of José Gaos*. Translated from Italian by Peter Cocozzella. Edited by Myra Moss. Introduction by Giovanni Gullace. A volume in **Values in Italian Philosophy**

49. Laura Duhan Kaplan and Laurence F. Bove, Editors, *Philosophical Perspectives on Power and Domination: Theories and Practices*. A volume in **Philosophy of Peace**

50. Gregory F. Mellema, *Collective Responsibility*

51. Josef Seifert, *What Is Life? The Originality, Irreducibility, and Value of Life*. A volume in **Central-European Value Studies**

52. William Gerber, *Anatomy of What We Value Most*

53. Armando Molina, *Our Ways: Values and Character*, Edited by Rem B. Edwards. A volume in **Hartman Institute Axiology Studies**

54. Kathleen J. Wininger, *Nietzsche's Reclamation of Philosophy*. A volume in **Central-European Value Studies**

55. Thomas Magnell, Editor, *Explorations of Value*

56. HPP (Hennie) Lötter, *Injustice, Violence, and Peace: The Case of South Africa*. A volume in **Philosophy of Peace**

57. Lennart Nordenfelt, *Talking About Health: A Philosophical Dialogue*. A volume in **Nordic Value Studies**

58. Jon Mills and Janusz A. Polanowski, *The Ontology of Prejudice*. A volume in **Philosophy and Psychology**

59. Leena Vilkka, *The Intrinsic Value of Nature*

60. Palmer Talbutt, Jr., Rough Dialectics: *Sorokin's Philosophy of Value*, with contributions by Lawrence T. Nichols and Pitirim A. Sorokin

61. C. L. Sheng, *A Utilitarian General Theory of Value*

62. George David Miller, *Negotiating Toward Truth: The Extinction of Teachers and Students*. Epilogue by Mark Roelof Eleveld. A volume in **Philosophy of Education**

63. William Gerber, *Love, Poetry, and Immortality: Luminous Insights of the World's Great Thinkers*

64. Dane R. Gordon, Editor, *Philosophy in Post-Communist Europe*. A volume in **Post-Communist European Thought**

65. Dane R. Gordon and Józef Niznik, Editors, *Criticism and Defense of Rationality in Contemporary Philosophy*. A volume in **Post-Communist European Thought**

66. John R. Shook, *Pragmatism: An Annotated Bibliography, 1898-1940*. With contributions by E. Paul Colella, Lesley Friedman, Frank X. Ryan, and Ignas K. Skrupskelis

67. Lansana Keita, *The Human Project and the Temptations of Science*

68. Michael M. Kazanjian, *Phenomenology and Education: Cosmology, Co-Being, and Core Curriculum*. A volume in **Philosophy of Education**

69. James W. Vice, *The Reopening of the American Mind: On Skepticism and Constitutionalism*

70. Sarah Bishop Merrill, *Defining Personhood: Toward the Ethics of Quality in Clinical Care*

71. Dane R. Gordon, *Philosophy and Vision*

72. Alan Milchman and Alan Rosenberg, Editors, *Postmodernism and the Holocaust*. A volume in **Holocaust and Genocide Studies**

73. Peter A. Redpath, *Masquerade of the Dream Walkers: Prophetic Theology from the Cartesians to Hegel*. A volume in **Studies in the History of Western Philosophy**

74. Malcolm D. Evans, *Whitehead and Philosophy of Education: The Seamless Coat of Learning*. A volume in **Philosophy of Education**

75. Warren E. Steinkraus, *Taking Religious Claims Seriously: A Philosophy of Religion*, Edited by Michael H. Mitias. A volume in **Universal Justice**

76. Thomas Magnell, Editor, *Values and Education*

77. Kenneth A. Bryson, *Persons and Immortality*. A volume in **Natural Law Studies**

78. Steven V. Hicks, *International Law and the Possibility of a Just World Order: An Essay on Hegel's Universalism*. A volume in **Universal Justice**

79. E. F. Kaelin, *Texts on Texts and Textuality: A Phenomenology of Literary Art*, Edited by Ellen J. Burns

80. Amihud Gilead, *Saving Possibilities: A Study in Philosophical Psychology*. A volume in Philosophy and Psychology

81. André Mineau, *The Making of the Holocaust: Ideology and Ethics in the Systems Perspective*. A volume in **Holocaust and Genocide Studies**

82. Howard P. Kainz, *Politically Incorrect Dialogues: Topics Not Discussed in Polite Circles*

83. Veikko Launis, Juhani Pietarinen, and Juha Räikkä, Editors, *Genes and Morality: New Essays*. A volume in **Nordic Value Studies**

84. Steven Schroeder, *The Metaphysics of Cooperation: A Study of F. D. Maurice*

85. Caroline Joan ("Kay") S. Picart, *Thomas Mann and Friedrich Nietzsche: Eroticism, Death, Music, and Laughter*. A volume in **Central-European Value Studies**

86. G. John M. Abbarno, Editor, *The Ethics of Homelessness: Philosophical Perspectives*

87. James Giles, Editor, *French Existentialism: Consciousness, Ethics, and Relations with Others*. A volume in **Nordic Value Studies**

88. Deane Curtin and Robert Litke, Editors, *Institutional Violence*. A volume in **Philosophy of Peace**

89. Yuval Lurie, *Cultural Beings: Reading the Philosophers of Genesis*

90. Sandra A. Wawrytko, Editor, *The Problem of Evil: An Intercultural Exploration.* A volume in **Philosophy and Psychology**

91. Gary J. Acquaviva, *Values, Violence, and Our Future*. A volume in **Hartman Institute Axiology Studies**

92. Michael R. Rhodes, *Coercion: A Nonevaluative Approach*

93. Jacques Kriel, *Matter, Mind, and Medicine: Transforming the Clinical Method*

94. Haim Gordon, *Dwelling Poetically: Educational Challenges in Heidegger's Thinking on Poetry*. A volume in **Philosophy of Education**

95. Ludwig Grünberg, *The Mystery of Values: Studies in Axiology*, Edited by Cornelia Grünberg and Laura Grünberg

96. Gerhold K. Becker, Editor, *The Moral Status of Persons: Perspectives on Bioethics*. A volume in **Studies in Applied Ethics**

97. Roxanne Claire Farrar, *Sartrean Dialectics: A Method for Critical Discourse on Aesthetic Experience*

98. Ugo Spirito, *Memoirs of the Twentieth Century*. Translated from Italian and Edited by Anthony G. Costantini. A volume in **Values in Italian Philosophy**

99. Steven Schroeder, *Between Freedom and Necessity: An Essay on the Place of Value*

100. Foster N. Walker, *Enjoyment and the Activity of Mind: Dialogues on Whitehead and Education*. A volume in **Philosophy of Education**

101. Avi Sagi, Kierkegaard, *Religion, and Existence: The Voyage of the Self*. Translated from Hebrew by Batya Stein

102. Bennie R. Crockett, Jr., Editor, *Addresses of the Mississippi Philosophical Association*. A volume in **Histories and Addresses of Philosophical Societies**

103. Paul van Dijk, *Anthropology in the Age of Technology: The Philosophical Contribution of Günther Anders*

104. Giambattista Vico, *Universal Right.* Translated from Latin and edited by Giorgio Pinton and Margaret Diehl. A volume in **Values in Italian Philosophy**

105. Judith Presler and Sally J. Scholz, Editors, *Peacemaking: Lessons from the Past, Visions for the Future*. A volume in **Philosophy of Peace**

106. Dennis Bonnette, *Origin of the Human Species*. A volume in **Studies in the History of Western Philosophy**

107. Phyllis Chiasson, *Peirce's Pragmatism: The Design for Thinking.* A volume in **Studies in Pragmatism and Values**

108. Dan Stone, Editor, *Theoretical Interpretations of the Hol*ocaust. A volume in **Holocaust and Genocide Studies**

109. Raymond Angelo Belliotti, *What Is the Meaning of Human Life?*

110. Lennart Nordenfelt, *Health, Science, and Ordinary Language*, with Contributions by George Khushf and K. W. M. Fulford

111. Daryl Koehn, *Local Insights, Global Ethics for Business*. A volume in **Studies in Applied Ethics**

112. Matti Häyry and Tuija Takala, Editors, *The Future of Value Inquiry*. A volume in **Nordic Value Studies**

113. Conrad P. Pritscher, *Quantum Learning: Beyond Duality*

114. Thomas M. Dicken and Rem B. Edwards, *Dialogues on Values and Centers of Value: Old Friends, New Thoughts*. A volume in **Hartman Institute Axiology Studies**

115. Rem B. Edwards, *What Caused the Big Bang?* A volume in **Philosophy and Religion**

116. Jon Mills, Editor, *A Pedagogy of Becoming*. A volume in **Philosophy of Education**

117. Robert T. Radford, *Cicero: A Study in the Origins of Republican Philosophy*. A volume in **Studies in the History of Western Philosophy**

118. Arleen L. F. Salles and María Julia Bertomeu, Editors, *Bioethics: Latin American Perspectives*. A volume in **Philosophy in Latin America**

119. Nicola Abbagnano, *The Human Project: The Year 2000*, with an Interview by Guiseppe Grieco. Translated from Italian by Bruno Martini and Nino Langiulli. Edited with an introduction by Nino Langiulli. A volume in **Studies in the History of Western Philosophy**

120. Daniel M. Haybron, Editor, *Earth's Abominations: Philosophical Studies of Evil*. A volume in **Personalist Studies**

121. Anna T. Challenger, *Philosophy and Art in Gurdjieff's* Beelzebub*: A Modern Sufi Odyssey*

122. George David Miller, *Peace, Value, and Wisdom: The Educational Philosophy of Daisaku Ikeda*. A volume in **Daisaku Ikeda Studies**

123. Haim Gordon and Rivca Gordon, *Sophistry and Twentieth-Century Art*

124. Thomas O. Buford and Harold H. Oliver, Editors *Personalism Revisited: Its Proponents and Critics*. A volume in **Histories and Addresses of Philosophical Societies**

125. Avi Sagi, *Albert Camus and the Philosophy of the Absurd*. Translated from Hebrew by Batya Stein

126. Robert S. Hartman, *The Knowledge of Good: Critique of Axiological Reason*. Expanded translation from the Spanish by Robert S. Hartman. Edited by Arthur R. Ellis and Rem B. Edwards.A volume in **Hartman Institute Axiology Studies**

127. Alison Bailey and Paula J. Smithka, Editors. *Community, Diversity, and Difference: Implications for Peace.* A volume in **Philosophy of Peace**

128. Oscar Vilarroya, *The Dissolution of Mind: A Fable of How Experience Gives Rise to Cognition*. A volume in **Cognitive Science**

129. Paul Custodio Bube and Jeffery Geller, Editors, *Conversations with Pragmatism: A Multi-Disciplinary Study*. A volume in **Studies in Pragmatism and Values**

130. Richard Rumana, *Richard Rorty: An Annotated Bibliography of Secondary Literature*. A volume in **Studies in Pragmatism and Values**

131. Stephen Schneck, Editor, *Max Scheler's Acting Persons: New Perspectives* A volume in **Personalist Studies**

132. Michael Kazanjian, *Learning Values Lifelong: From Inert Ideas to Wholes*. A volume in **Philosophy of Education**

133. Rudolph Alexander Kofi Cain, Alain Leroy Locke: *Race, Culture, and the Education of African American Adults*. A volume in **African American Philosophy**

134. Werner Krieglstein, *Compassion: A New Philosophy of the Other*

135. Robert N. Fisher, Daniel T. Primozic, Peter A. Day, and Joel A. Thompson, Editors, *Suffering, Death, and Identity*. A volume in **Personalist Studies**

136. Steven Schroeder, *Touching Philosophy, Sounding Religion, Placing Education*. A volume in **Philosophy of Education**

137. Guy DeBrock, *Process Pragmatism: Essays on a Quiet Philosophical Revolution*. A volume in **Studies in Pragmatism and Values**

138. Lennart Nordenfelt and Per-Erik Liss, Editors, *Dimensions of Health and Health Promotion*

139. Amihud Gilead, *Singularity and Other Possibilities: Panenmentalist Novelties*

140. Samantha Mei-che Pang, *Nursing Ethics in Modern China: Conflicting Values and Competing Role Requirements*. A volume in **Studies in Applied Ethics**

141. Christine M. Koggel, Allannah Furlong, and Charles Levin, Editors, *Confidential Relationships: Psychoanalytic, Ethical, and Legal Contexts*. A volume in **Philosophy and Psychology**

142. Peter A. Redpath, Editor, *A Thomistic Tapestry: Essays in Memory of Étienne Gilson*. A volume in **Gilson Studies**

143. Deane-Peter Baker and Patrick Maxwell, Editors, *Explorations in Contemporary Continental Philosophy of Religion*. A volume in **Philosophy and Religion**

144. Matti Häyry and Tuija Takala, Editors, *Scratching the Surface of Bioethics*. A volume in **Values in Bioethics**

145. Leonidas Donskis, *Forms of Hatred: The Troubled Imagination in Modern Philosophy and Literature*

146. Andreea Deciu Ritivoi, Editor, *Interpretation and Its Objects: Studies in the Philosophy of Michael Krausz*

147. Herman Stark, *A Fierce Little Tragedy: Thought, Passion, and Self-Formation in the Philosophy Classroom*. A volume in **Philosophy of Education**

148. William Gay and Tatiana Alekseeva, Editors, *Democracy and the Quest for Justice: Russian and American Perspectives*. A volume in **Contemporary Russian Philosophy**

149. Xunwu Chen, *Being and Authenticity*

150. Hugh P. McDonald, *Radical Axiology: A First Philosophy of Values*

151. Dane R. Gordon and David C. Durst, Editors, *Civil Society in Southeast Europe*. A volume in **Post-Communist European Thought**

152. John Ryder and Emil Višňovský, Editors, *Pragmatism and Values: The Central European Pragmatist Forum, Volume One*. A volume in **Studies in Pragmatism and Values**

153. Messay Kebede, *Africa's Quest for a Philosophy of Decolonization*

154. Steven M. Rosen, *Dimensions of* Apeiron: *A Topological Phenomenology of Space, Time, and Individuation*. A volume in **Philosophy and Psychology**

155. Albert A. Anderson, Steven V. Hicks, and Lech Witkowski, Editors, *Mythos and Logos: How to Regain the Love of Wisdom*. A volume in **Universal Justice**

156. John Ryder and Krystyna Wilkoszewska, Editors, *Deconstruction and Reconstruction: The Central European Pragmatist Forum, Volume Two*. A volume in **Studies in Pragmatism and Values**

157. Javier Muguerza, *Ethics and Perplexity: Toward a Critique of Dialogical Reason*. Translated from the Spanish by Jody L. Doran. Edited by John R. Welch. A volume in **Philosophy in Spain**

158. Gregory F. Mellema, *The Expectations of Morality*

159. Robert Ginsberg, *The Aesthetics of Ruins*

160. Stan van Hooft, *Life, Death, and Subjectivity: Moral Sources in Bioethics* A volume in **Values in Bioethics**

161. André Mineau, *Operation Barbarossa: Ideology and Ethics Against Human Dignity*

162. Arthur Efron, *Expriencing Tess of the D'Urbervilles: A Deweyan Account.* A volume in **Studies in Pragmatism and Values**

163. Reyes Mate, *Memory of the West: The Contemporaneity of Forgotten Jewish Thinkers.* Translated from the Spanish by Anne Day Dewey. Edited by John R. Welch. A volume in **Philosophy in Spain**

164. Nancy Nyquist Potter, Editor, *Putting Peace into Practice: Evaluating Policy on Local and Global Levels.* A volume in **Philosophy of Peace**

165. Matti Häyry, Tuija Takala, and Peter Herissone-Kelly, Editors, *Bioethics and Social Reality.* A volume in **Values in Bioethics**

166. Maureen Sie, *Justifying Blame: Why Free Will Matters and Why it Does Not*. A volume in **Studies in Applied Ethics**

167. Leszek Koczanowicz and Beth J. Singer, Editors, *Democracy and the Post-Totalitarian Experience*. A volume in **Studies in Pragmatism and Values**

168. Michael W. Riley, *Plato's* Cratylus: *Argument, Form, and Structure*. A volume in **Studies in the History of Western Philosophy**

169. Leon Pomeroy, *The New Science of Axiological Psychology*. Edited by Rem B. Edwards. A volume in **Hartman Institute Axiology Studies**

170. Eric Wolf Fried, *Inwardness and Morality*

171. Sami Pihlstrom, *Pragmatic Moral Realism: A Transcendental Defense.* A volume in Studies in **Pragmatism and Values**

172. Charles C. Hinkley II, *Moral Conflicts of Organ Retrieval: A Case for Constructive Pluralism.* A volume in **Values in Bioethics**

173. Gábor Forrai and George Kampis, Editors, *Intentionality: Past and Future.* A volume in **Cognitive Science**

174. Dixie Lee Harris, *Encounters in My Travels: Thoughts Along the Way.* A volume in **Lived Values:Valued Lives**

175. Lynda Burns, Editor, *Feminist Alliances.* A volume in **Philosophy and Women**

176. George Allan and Malcolm D. Evans, *A Different Three Rs for Education.* A volume in **Philosophy of Education**

177. Robert A. Delfino, Editor, *What are We to Understand Gracia to Mean?: Realist Challenges to Metaphysical Neutralism.* A volume in **Gilson Studies**

178. Constantin V. Ponomareff and Kenneth A. Bryson, *The Curve of the Sacred: An Exploration of Human Spirituality.* A volume in **Philosophy and Religion**

179. John Ryder, Gert Rüdiger Wegmarshaus, Editors, *Education for a Democratic Society: Central European Pragmatist Forum, Volume Three.* A volume in **Studies in Pragmatism and Values**

180. Florencia Luna, *Bioethics and Vulnerability: A Latin American View.* A volume in **Values in Bioethics**

181. John Kultgen and Mary Lenzi, Editors, *Problems for Demo*cracy. A volume in **Philosophy of Peace**

182. David Boersema and Katy Gray Brown, Editors, *Spiritual and Political Dimensions of Nonviolence and Peace.* A volume in **Philosophy of Peace**

183. Daniel P. Thero, *Understanding Moral Weakness.* A volume in **Studies in the History of Western Philosophy**

184. Scott Gelfand and John R. Shook, Editors, *Ectogenesis: Artificial Womb Technology and the Future of Human Reproduction.* A volume in **Values in Bioethics**

185. Piotr Jaroszyński, *Science in Culture*. A volume in **Gilson Studies**

186. Matti Häyry, Tuija Takala, Peter Herissone-Kelly, Editors, *Ethics in Biomedical Research: International Perspectives*. A volume in **Values in Bioethics**

ഗ്രീൻ ബുക്സ്

ഒരു പ്രവാസിയുടെ ഡയറിക്കുറിപ്പുകൾ

ഡോ. റെജി ഡി. നായർ

അധ്യാപകൻ, സാമ്പത്തിക ശാസ്ത്രജ്ഞൻ,
പരിസ്ഥിതി പ്രവർത്തകൻ.

വിദ്യാഭ്യാസം: 1997-ൽ കേരള യൂണിവേഴ്സിറ്റിയിൽ നിന്നും ഇക്കണോമിക്സിൽ ഡോക്ടറേറ്റ്. എത്യോപ്യയിലെ അർബാമിഞ്ച് യൂണിവേഴ്സിറ്റി, സൗദി അറേബ്യയിലെ റോയൽ കമ്മീഷൻ കോളേജ് എന്നിവിടങ്ങളിൽ അസ്സിസ്റ്റന്റ് പ്രൊഫസറായി സേവനമനുഷ്ഠിച്ചിട്ടുണ്ട്. 'ഗ്രാമത്തിൽ ഒരു അവധിക്കാലം' എന്ന കുട്ടികളുടെ നോവലിന്റെ ഇംഗ്ലീഷ് പരിഭാഷ ഗുഡ് ടൈംസ് ബുക്സ്, ന്യൂഡൽഹി 'സമ്മർവെക്കേഷൻ' എന്നീ കൃതികൾ പ്രസിദ്ധീകരിച്ചിട്ടുണ്ട്.

ഭാരതീയ ദളിത് സാഹിത്യ അക്കാദമിയുടെ ഡോ. അംബേദ്ക്കർ ഫെലോഷിപ്പ് അവാർഡ് (2006), ഗീതാഞ്ജലി സാംസ്കാരിക വേദിയുടെ പ്രവാസ സാഹിത്യ പുരസ്കാരം (2016) തുടങ്ങിയ പുരസ്കാരങ്ങൾ.
സാന്ദ്രസ്പർശം (2017) എന്ന ഹ്രസ്വചിത്രത്തിന്റെ നിർമ്മാതാവ് കൂടിയാണ് റെജി.
ഇപ്പോൾ യു.എ.ഇയിൽ ഹയർ കോളേജ് ഓഫ് ടെക്നോളജിയിൽ അധ്യാപകനായി ജോലി ചെയ്യുന്നു.
നിരവധി കൃതികളുടെയും രചയിതാവാണ്.